AUTOBIOGRAPHY OF A HAPPY SCIENTIST

Tips and Tricks for Surviving in Science

KEVIN THOMAS MORGAN

After my final scientific lecture, "*Paradigm Shifts In Biology*," several young pathologists in the audience came up to ask me, "What would you do if you were just starting your career in the biological sciences?"

I replied without hesitation,

"Choose a trade, master it, then study mathematics and multiple branches of engineering as cells are exquisite engineers."

It's what they do!

Science Survival Tip No. 1: *It is not the strongest of the species that survives, nor the most intelligent. It is the one that is most adaptable to change.*

— Charles Darwin

FOREWORD

If you want to enjoy a career in science, my advice is that you first master a trade. A trade that is needed in areas of science that interest you, and one that will permit you to earn a living if the grants don't come in.

I have endeavored in this little book to provide useful insights, concerning a career in science. I have changed the names of some people for the sake of discretion. I never intended, consciously at least, to become a research scientist, but I did. Life just carried me along. There are certain tricks to enjoying a life in science, many of which are presented here, based on personal experience, and are contained within some excellent publications recommended throughout the narrative. Having a marketable trade is an excellent place to start.

My trade is pathology or the study of the nature of disease, much of it looking down a microscope. For the first ten years of my career, I was hired as a pathologist, not as a researcher. Once I had twenty or more publications under my belt, in decent scientific journals, the job offers were research-based.

This is the method that has worked for me. I don't like being unable to provide for myself and my family. Furthermore, I don't like doing jobs I don't like.

Been there, done that!
Think about it.

Science Survival Tip No. 2: *Become a Jack of all trades, master of ONE, the trade that will permit you to earn a living.*

ALSO BY KEVIN THOMAS MORGAN

We Can't Eat Grass

How to Train for Aging

The True Story of Plantar Fasciitis

Plantar Fasciitis Has The Wrong Name:

Find Peace of Mind in the Pool

Pain, Good Friend, Bad Master

A Tiny House Fixed My Retirement Cash Flow

Aortic Disease From The Patient's Perspective

Surgery Recovery Guide

Body Meditation for Optimal Movement

Changing the Way You Move

FitOldDog's Plantar Fasciitis Treatment

INTRODUCTION

Science Survival Tip No. 3: *Don't panic and remember to bring a large bath towel and a SEP field generator.*

It was raining the day I was born, in June 1943. It was raining incendiary bombs, dropped by Nazi planes during World War II. I became one of the many boys playing on the bombsites of Britain.

You might be thinking, "Poor boys, playing on rubble and fallen buildings."

You'd be wrong.

We had a blast. Kids of all ages, having rock fights, hiding behind half-demolished walls, searching for secret cellars and hidden bomb shelters. Each boy's prized possession was his personal crowbar, that was used to bring down walls, find secret passages, looking for bodies or gas masks. Fortunately, I never did find any skeletons, but one kid proudly showed us a gas mask in near-mint condition, that he had uncovered in the rubble.

We had fun playing on the remnants of terror and destruction. We were kids. It was all we knew.

I remember a clock. An old, wood-framed, windup time-piece. It was sitting on a high mantelpiece jutting out into space, four floors above the remains of a stricken row house. The rest of the house was a pile of stone, bricks and rotting timbers. How had that clock survived? The bombsite was like a missing tooth in an otherwise healthy smile. The adjacent homes appeared unimpressed by the absence of their neighbor.

The other boys would try to knock that old clock down by throwing stones. Even when they used homemade catapults, it remained steadfast. *I expect it's still showing the time when the building fell during the war,* I thought. I never considered it wise to hurl rocks up there. Having been unable to climb up to see the time on that mysterious memento of days long past, I would dream of flying up there.

I have always wanted to fly. I still do. Not with a machine, more like Superman or Peter Pan, using special powers. Even now, in my 70s I try to think my body off the ground from time to time, with no success so far.

I know!

I am well beyond my allotted three score years and ten. Having spent many years studying fluid mechanics, I am well aware that unaided human flight is highly unlikely. I'm old enough to know better. Fortunately, I refuse to know better.

INTRODUCTION

Science Survival Tip No. 4: *Never give up trying to fly. I don't care how old you are.*

You can laugh at an old man's folly, but one day you might see me skimming across the treetops, and then you'll wish you'd tried to master flight before it was too late.

In my boyhood dreams, I floated up to that high mantlepiece. The time on the clock never changed: 22 minutes past four. Whether it was the morning or afternoon, I had no idea. I would place my ear close and hear a faint *tick, tock, tick, tock.* The clock was waiting for the owners to return and wind it up with an old brass key.

From time to time, I would search for that key amongst the rubble down below.

I think my sense of curiosity and adventure served me well during a 40-year career as a research scientist. This career is presented as a dry list of scientific publications in the appendix. However, my life as a scientist involved suspense, discovery, human science parasites and mimics, mind-opening studies of physics and French, battles with journal editors and science managers.

The work itself involved sitting at microscopes for hundreds of hours, followed by grinding number crunching. However, this was totally offset by delightful Friday evening discussions in the pub, after work, or over business lunches. There was a flight to Mexico City from North Carolina to have lunch with a scary Mexican businessman, to facilitate the rescue of a truck that had been impounded by Mexican customs for three months. This truck contained $500,000 worth of laboratory equipment, for an experiment on the toxicity of Mexico City air. That otherworldly conversation rescued our equipment while reminding me of *The Godfather*. The truck was never mentioned during the meal. He left me with the check, a clear message to me of who was in control. Our truck was rolling back toward Mexico City two days later.

Do you still think science is a straightforward process?

My science career took me to the Great Wall of China, and during that trip I got to work out with the Chinese Kung Fu team. Martial arts was another of my obsessive hobbies that taught me a great deal about human biomechanics. Of course, they kicked my butt, but what a treat.

I have had many weird adventures, such as learning how to make sheep's milk, minus vitamin B_1, from its individual components in an old washing machine. That was when I learned about double emulsions and how important it is to add ingredients in the correct order. I looked directly down a rifle barrel that was pointed in my face by a member of the Chinese Red Guard. He wasn't joking. My only thought at the time was, *Boy, it's dark down there.*

Then there was an amusing discussion with six customs agents at Heathrow Airport over a large bag of decalcified rat heads. I handed over the bag of rat heads, along with the correct paperwork. The customs agent, a nice young man, asked what it was about. I explained that I was on my way to run a one day course on the pathology of the nose at Cambridge University. He looked worried, and said he had to consult his boss.

A small group of customs officials congregated about 20 feet away. They would glance in my direction from time to time before returning to their debate, looking concerned. Clearly, they did not know what to do. After about 15 minutes, the original agent broke away from the now smiling group. He leaned toward me to whisper in my ear:

"Sir, if you don't tell anyone, nor will we." He then added, "Have fun in Cambridge. You're free to proceed."

And on I went.

Go figure.

I had innumerable other fascinating adventures. The stuff of science in the trenches that is never mentioned in the articles they lead to. These experiences and the lessons they provide are the subject of this book. The life of a research scientist, from start to an undefined finish.

Finally, I'd like to mention the greatest compliment I have ever received. I did not receive the Nobel Prize though the money would be nice or any other major awards. But I got to see a raft of graduate students and post-doctoral fellows under my tutelage move on to successful careers. In fact, it was one of these who gave me my last job in science. That was a compliment for sure, and a wonderful job it was.

This greatest scientific award of my career was provided by a young theoretical physicist named Bahman. This delightful man worked in our fluid mechanics program for a while. I had not seen him for nearly 10 years before he sought me out at a scientific meeting.

"Kevin, I'm so glad to see you." He said. "I've wanted to thank you for a long time."

"Hi Bahman. Thank me for what?"

"For teaching me how to question the obvious!"

That one comment was more important to me than any fancy letters after my name.

Science Survival Tip No. 5: *Learn how to question the obvious, which is almost never obvious.*

For instance, it's obvious that we live in a 3D Cartesian world with the exception of the weirdness of spacetime. Right? That said, I saw something that should not surprise anyone who has studied a little string theory.

You have, haven't you? But is it wrong?

After spending about 10 years trying to understand airflow in the nose of the rat and other species including people, using computational fluid dynamics software, I became consumed with mathematics, especially geometry. Then our mathematician, Julia Kimbell aka Julie, said she had ordered two movies for the group to watch.

Off we went to a conference room and cranked up the old 16mm projector. Have you ever heard the clatter they make? Boy, those movies were a wonderful mind opener. It pays to have an active imagination with a mathematician available.

Go then, there are other worlds than these. *Jake Chambers*. Fantasy novels provide a great way to work on your imagination.

You can imagine the five of us, including two pathologists, a statistician, a sculptor and a mathematician, sitting in the dark waiting for the show. It starts up as the film spools, and the movie plays on the white screen.

The first movie, *Flatland*, was about a little square living in a 2D world. Julie selected it to soften us up, to open up our minds to whatever revelation was about to come. And a revelation it was, for me at least.

Flatland

A little square was busying himself by going about the activities of the day, chatting to circles, rhomboids, trapezoids, triangles and all the other two-dimensional creatures that live in Flatland. All seemed fine, until one day the little square encountered a new, strange circle.

This circle was odd. It changed size from small to large, and back again. The little square was puzzled, and said to the circle, "What kind of a circle are you? You change size all the time." The circle replied, "I'm not a circle, I'm a sphere intersecting your world. As I move from one side to the other, I appear as a circle of variable diameter. I'm not a circle at all, I'm a sphere, and I live in a three-dimensional world, which you can't imagine."

The square found this hard to believe, if not a little insulting. There was a long discussion as the sphere, somewhat arrogantly, educated the square on the nature of 3D space. He told the square that the square's relatives in 3D space were known as cubes. The square was a bright chap, and he cottoned on quickly, and said, "I get it, Mr. Sphere, thanks so much. So you live in three dimensions?" "Yes!" Replied the sphere, pompously.

Obviously, he was a more-educated shape than the square, and he was pleased to share his wisdom.

The square thought for a while, and then said excitedly, "But, Mr. Sphere, if there are really three dimensions, as you've explained, then there could be four, or five, or even more?"

The sphere stated with conviction that this was not possible, but the square wanted to know how he knew that. The square had only just been alerted to the third dimension. Maybe, one day the sphere might encounter a four-dimensional creature intersecting his world, and then the square was concerned that the sphere might be too closed minded to recognize it. Of course, being polite, the square refrained from making this remark.

The sphere was a proud creature, and he continued to deny the existence of a four-dimensional world.

The square thanked him for his time, realizing that further discussion was a waste of time. As he moved away, the square said quietly to himself, "I bet there is a four-dimensional world, and maybe a five and a six and even more." Off he went, dreaming of these wonderful places and how he might explore them one day.

I was intrigued and determined to never be closed minded like the arrogant sphere. I loved the open-minded attitude of the little square. Then came the reason for this flight of fancy.

The Four Dimensional Cube

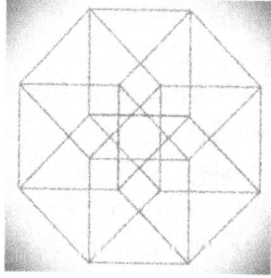

We proceeded to watch the second movie, entitled, *The Tesseract*. A grainy image of a dot appeared on the screen. The movie was old and scratched, presenting a computer simulation of various shapes that stepped from one dimension up to the next. Each set of images was built on the one before, in a logical and educational progression. Perfect for the little square, but he wasn't in the room, unfortunately. He would have loved it.

It was explained that the dot lived in zero-dimensional space. A limited place, indeed. Then a line appeared, a one-dimensional object. The movie showed the line being transected by the dot, a moving point. A one dimensional line was being bisected into two pieces, as the point moved along it. A zero-dimensional object can cut a one-dimensional object into two.

Then a two-dimensional square appeared on the screen. You have probably guessed by now what will happen next, as the two-dimensional square is bisected by the one-dimensional line, that moves across it, dividing the square into two rectangles.

Then, on the two dimensional projector screen, a drawing of a three-dimensional cube appears. It's being bisected by a moving two-dimensional square. A three dimensional object is being divided into two parts by a moving two dimensional object.

Got it. Easy.

I was mesmerized by the magic that occurred next. A mess of lines appeared on the screen. The commentator explained that it was a four-dimensional cube, a tesseract. The movie proceeded to bisect the four-dimensional cube with the three-dimensional cube, all on a two dimensional screen.

You can hear the sphere saying, "That's a lot of nonsense," while the little square and myself sit enchanted.

After the second movie finished, most of the group wandered away, but they didn't seem too impressed. I asked Julie if I could watch the tesseract again. *No problem!* Off she went, leaving me to watch the movie to my hearts content, which I did. Dozens and dozens of times.

Here comes the magic.

Sitting alone in the dark, something incredible happened. After about 50 repeats of the movie I started to have visual flashes of four-dimensional space. I could clearly see the four-dimensional tesseract being bisected by a three-dimensional cube. Without the movement it would have been invisible, I'm sure. This was even more exciting to me than throwing methanol on a coal fire, but I'm getting ahead of myself.

I was, and still am, convinced that we actually live in an n-dimensional universe, with $n \geq 4$.

Seeing is believing, right?

Well, maybe not. I'm not one for taking drugs, except for an occasional glass of wine, but a good friend of mine told me an interesting story about his one and only LSD experience:

"Kevin, I had a weird experience last night. I took some LSD and was sitting in front of the fire. You know that big wood-framed clock on my mantlepiece, that chimes from time to time? Well, I was sitting there wondering what would happen, as I've never tried LSD before, and I won't again.

I was relaxing when something dripped off of the mantlepiece into the fire grate. I noticed then that the clock was melting. In fact, it proceeded to completely melt, dripping down onto the grate to form a multicolored puddle. It seemed like a normal thing to happen, until I woke up in the morning and the clock was back on the mantlepiece. I couldn't believe it. I was sure it would be a puddle on the grate."

He never did try LSD again, he told me, but it made me wonder about that expression, *seeing is believing*.

The tesseract movie that intrigued me so was made a long time ago, using crude technology by today's standards. Now you can go online and see sophisticated virtual reality renditions of n-dimensional space. You have to think like the open-minded little square from Flatland if you hope to glimpse four-dimensional space.

Or was I imagining things as a product of wishful thinking? Is string theory a load of hot air, is the world flat after all?

That you will have to decide for yourself.

Tick tock, tick tock.

This book will make you wonder about such things. That's what science is about, wondering about things.

PROLOGUE

Highly recommended treatise on the impact of technology on epistemology, meaning technology can impact your state of knowledge, and thus your wisdom.

Science Survival Tip No. 6: *The key currency of science, most of the time, is publications in peer reviewed scientific journals, so publish, publish, publish.*

I love to read. Reading rescued my career, as you'll see below. I considered putting a recommended reading list at the end of the book. That seemed boring, and no one would read it, so I created mockups of books I recommend that are plastered throughout this book.

Enjoy.

Full references to articles derived from studies referred to in the narrative are indicated by their numbers in the appendix.

MABEL AND MORE

Science Survival Tip No. 7: *Observe and question the obvious before attempting to solve.*

Winter 1970

It's the end of a long day tending to livestock. I am ready for my dinner! I had been working as a country veterinarian on farms in Southwest England for three years.

Tonight, an old farmer and I are in a cowshed by the light of a single, grimy light bulb. A set of old buildings, straw, hay, mud and hard work surrounded us. Cobwebs hang from the rafters as 15 dairy cows, Jerseys, are chewing the cud in the evening gloom.

Delightful animals that are as gentle and affectionate as you can imagine.

This farmer is just getting by, that's for sure. He's looking at me with old, worried eyes, from a heavily wrinkled face. This man has seen, and is still living, hard times, but he cares about his livestock. I came to treat one of his cows for mastitis.

"I guess you can't do nowt for old Mabel, young man?" he says, as I write the ticket for the bill

I was not one of the owners of the veterinary practice. I was a hired hand, an assistant, as opposed to a partner in the business. I could not waste time or other resources.

"What's up with Mabel?" I replied. Which one was Mabel, I had no idea.

"She can't eat, and she's losing weight. Mabel's a family pet. She gave up giving milk years ago. We like to keep her around, you know."

I took a look at Mabel, a kind soul. She had managed to break her mandibular symphysis. That's where the two halves of the lower jaw join together at the front. Fingers of bone on each side interdigitate to create a single solid structure. Imagine linking your hands by clasping your fingers. Left, right, left, right. That's how the mandibular symphysis works, combined with strong bands of dense connective tissue.

This nice old farmer doesn't have a lot of money. He doesn't want to send her off to slaughter either, and she's a family friend. I racked my brains for an affordable – even free - solution. Major surgery, with plates and screws, would cost a lot, even if it was possible with an old dairy cow.

It was the end of a long day.

I'd earned my living and am about to head home. Then a thought strikes me! I can do this job, on my own time. A length of strong, absorbable suture material, cat gut - about tuppence, local anesthetic - another two pennies, and a drug to keep her calm - a tanner, at most. I'd see if I could pull that jaw back together. It'll be a tough job, but I'll give it a go.

I cleaned out the joint and bound the two halves together with a figure-of-eight suture knot. Imagine someone passing string around the palms of your interdigitated hands in a figure of eight. Over the top of the right hand. Down and under your left. Then over the top of your left, then down and under your right. Several turns to pull your hands, and thus your opposing fingers, together. The only difference is the rigid nature of jaw bones and the unpredictable behavior of a cow's head,

Could I pass the cat gut under the jaw bones without damaging any nerves or crushing blood vessels?

That's what we did, the old farmer and I. He distracted Mabel, and I performed the surgery, crude as it was, and it only took an hour!

I came back two weeks later. The farmer was delighted in his subdued, Somerset-Farmer way. Mabel was happy, putting on weight again. The case of mastitis had cleared up too. We'd saved that old cow for less than a bob (about $5 today), with a little thought and some surgical experience.

I'd observed the problem: What did this observation process involve? Well, the cow's jaw, for sure, and her general state of health. Could Mabel handle the intended surgery. Then there was the farmer, his financial situation and concern for Mabel. There was also my relationship to the farmer and the veterinary practice, which led me to do this work inexpensively on my own time. Then came my surgical experience. Could I take on such a surgery? Observation needs to include the entire context of the situation, as it will for anyone who attempts to tackle the challenges of scientific research.

This also required questioning available approaches - A simple issue as the choices were limited to doing nothing, sending her to the experts at the local veterinary school, or working it out myself.

Solving Mabel's dilemma: In this case, saving the farmer's cow from the slaughterhouse, with the supplies in my car, a Morris Traveler. I loved that car, it was perfect for a farm veterinarian.

Finally, I had to affordably help that old farmer. Why do I consider affordability to be so important in science? Lab budgets are rarely unlimited and a good scientist dispenses their resources wisely.

These are the basics of the method I eventually used as a scientist for the next 40 years or so. In fact, I still am. I call it the OQS method:

- Observe and carefully define the problem.
- Question the obvious.
- Solve the problem, if you can.

Humans tend to rush toward solutions before fully understanding the context of a problem. This makes sense. Our primitive ancestors who ran when they saw rustling grass, assuming the presence of a tiger, were more likely to survive than those who stood around waiting to see if it was just the wind or actually a tiger. It paid to rush to the quickest solution. Run!

Avoiding tigers is not a good metaphor for the process of experimental design or data interpretation. Early in my career, I certainly erred in this direction a few times, but after about 30 years, I was able to resist the pressure to publish prematurely, whether it was due to personal enthusiasm, an overbearing manager or an impatient granting agency.

As a pathologist, I developed a trick for training my students in the art of patience during the observation process. It was a simple trick that never failed to irritate for the first few minutes. I recommend it for any scientists in your lab, including whether they are pathologists, biochemists or mathematicians. Everyone should get to know their tissues and cells of interest down the microscope, even theoretical physicists and statisticians.

My instructions went like this:

- Sit and relax at a microscope.
- Choose a microscope slide with a busy field, at 20x magnification, with lots of things going on.
- Set up a timer for 10 minutes.
- Look down the scope, without moving the slide, and start the timer.
- See what you see.

I would sit nearby and listen in. The response was always the same. *OK! I see this and this and that.* Two minutes: *"There's nothing else to see!"* Three minutes. Four minutes, *"Sigh!"* Five minutes, *"OMG, I didn't notice that ... "* The boredom vanished, transformed into a competition to see what else they might find. They were on the road to becoming accomplished morphologists.

What if they weren't pathologists, but molecular biologists? The same would apply. Looking at the data, within a chosen context. For the 20x field, the context of an observation is all that is present in the slide, combined with the observer's database (education, training, level of receptivity to new ideas, etc.). In the case of a molecular biologist, the context has to be constructed by reading the literature, and ideally creating a state diagram of the network of information surrounding the issue of interest.

Another observation training tool I used was a large color photograph of a field, with deciduous and evergreen trees in the distance. In the foreground were flowering plants, ferns and horsetails. Closer by were some mosses and lichens on a dead branch. If I could have found such a picture, I would have included some liverworts, my favorite bryophytes.

I have been walking and running on this trail for over 30 years, and this is the first liverwort I've seen. I suspect it was hanging out as spores until the climate changed.

I would ask students what they saw in the photo and they would look askance at me. What do you see in the picture above, I wonder? Some would say "green stuff." Others, a field with some trees in the distance. A few might say there were both evergreen and deciduous trees, some flowers and a few ferns, which was as good as it got.

Then I would explain what I could see.

The evolution of the plant kingdom, along with symbiosis, and the battle for survival. Furthermore, I imagined the underground fungal mycelia and countless billions of microscopic organisms that make up the biosphere. All working together for mutual survival or competing in a no holds barred battle for space and light.

Want to know more? Read this remarkable book, *The Hidden Life Of Trees*.

If you read this book you'll never look at the woods in the same way again. It's much more alive and integrated than I realized.

Science Survival Tip No. 8: *You only see what you know, so learn and learn some more, and you might see more than you can ever imagine.*

Start by observing, and mastering the context within which your idea or question is embedded. Then consider solutions and experiments to confirm or refute them.

About 20 years ago, I was sitting on a dataset derived from an experiment that cost in excess of $1 million. I looked at our data every which way, but I could not compile a coherent story. Something was missing, while the data were clearly indicating a pattern. A year went by, and the president of the institute would come by to ask how that particular paper was coming along. I would smile, and say, *Sorry Roger, we are still trying to work out what it all means.*

After two years, open hostility was generated by this reply, with encouragement to hurry up.

Not a chance, I couldn't make sense of these data.

Three years went by, the same result, then I had a chat with a statistician recommended to me by a friend. I explained the problem, and he replied, "Have you carefully considered all possible denominators?" Six months later the article went to press, and was deemed to be of considerable interest to researchers in the field of quantitative risk assessment. I did the right thing to bide my time until the appropriate denominator was identified.

I could do this because I was established in my field of research. A beginner would have a tough time with a pushy manager, unless they had a powerful supporter on their side. That statistician ended up as a co-author on the article, because he played a critical role in the final analyses and data interpretation.

Science Survival Tip No. 9: *Experience is the best teacher (but only when the experience isn't fatal).*

— Peter J Feibelman

WHERE DO IDEAS AND SOLUTIONS COME FROM?

Science Survival Tip No. 10: *Ideas come from other ideas, and to be seen as a genius, rather than a pirate, hide your sources.*

— David Kord Murray and Albert Einstein

Finding and defining problems is more important, and generally more difficult, than solving them. Once you know the problem you are tackling, you can set your course. We are trained in school to solve problems, but we are rarely asked to find or define problems. This is what scientists and other entrepreneurs do for a living.

For instance, in 1981, I looked down a microscope at a section of the nose of a rat that had been exposed to formaldehyde gas one day before it was killed and its nasal passages prepared for examination in a microscope. I noticed something odd, which led to a 10-year search for the solution to a question.

Science Survival Tip No. 11: *The answers you get depend upon the questions you ask.*

— Thomas Kuhn

I had found and defined an interesting and potentially important problem! Much of science is the result of an observation that leads to a question involving an interesting problem.

Think of the story of Sir Isaac Newton and the falling apple.

Fall, 1991, Sunday morning, at 8:00 a.m.

The phone is ringing.
"Kevin, it's Jim."
"What's up?"
"I know the answer. Fill the flow tank with dilute developer. Place an exposed sheet of X-ray film on the floor of the tank with something on top to hold it down and create a complex flow field over the film, and..."
"Got it, Jim!"

I hung up the phone, drove to the lab, set up the experiment that Jim proposed. I used a piece of metal to hold down the film, selected to create an interesting downstream flow field over the film, to include swirls and eddies. Within five minutes, I was treated to the most remarkable development pattern. It made complete sense of my observation in that slide of that poor rat exposed to formaldehyde gas 10 years earlier.

Yes! Ten years.

We were trying to find out why damage induced by inhaled formaldehyde in the nasal passages of rats was so site-specific [40]. In some areas, the nasal epithelium was destroyed with surgical precision, leaving adjacent tissue untouched.

It was not mucus flow patterns we had tested that [33].

It was not local biochemistry we had examined that [63].

It was not local blood flow we had explored that, but never got around to publishing the work because I took another job, and that happens. Don't sweat it, you can't finish everything.

The observation of surgically precise tissue damage was due to airflow patterns and the boundary layer structures they create close to the airway wall. You can read similar stories about tornadoes. One house destroyed, next door untouched, but we needed to work out how to prove it for reactive gases, not houses.

These inhaled gases reach the nasal lining in two ways. Convection, carried along in the bulk airflow, and diffusion out of the bulk flow, towards the nasal lining. Adjacent to the wall of the nose and the flow tank is an unstirred layer, of air or water, respectively, across which dissolved materials diffuse. The thinner the unstirred layer, the quicker these materials reach the wall.

In the case of the nose this gas delivery process will determine how quickly formaldehyde goes from the bulk, convective flow field, to the wall. The more formaldehyde that reaches the wall, the more the damage.

The dose makes the poison. – Paracelsus

In the flow tank, the same things occur, except in this case it's the diffusion of developer, or hydroquinone, that determines the rate of development of the exposed film. The more hydroquinone carried to the film by convection, combined with more rapid diffusion in areas with a thinner unstirred boundary layer, the more quickly the film clears. That is how the flow tank experiment revealed the remarkable precision of the delivery, with knife like sculpting of the development, by a complex fluid flow pattern over a solid surface [79].

Our work with that flow tank was consistent with the conclusion that the precise localization of lesions in the nose, induced by inhaled formaldehyde [79], is the result of the combination of regional bulk delivery rate and the structure of the air-phase boundary layer adjacent to the lining of the nose.

This sure got us excited after our 10 years of research to solve what seemed like a simple problem. This is how science is done, by dogged pursuit. Now we had a way to employ computational fluid dynamics, rather than rats, to estimate local gas dosimetry in multiple species, including humans [116].

This was all thanks to Jim's dedication and imagination. Jim was a slim, quietly spoken, man of average height in his 50s. He requested to work in our flow lab for his six-month sabbatical from his normal faculty position up north. I was delighted, given the fact that he was an expert in chemical and biomedical engineering.

Jim settled into the lab quickly, with its array of delicate equipment, from rheometers to oscilloscopes and even a large Faraday cage for high impedance, micro-electrode work. He was in his element.

Science Survival Tip No. 12: *Do not let everyone in your lab, especially theoreticians, as they might break your equipment. I'm not joking.*

Much of our research was done with a Plexiglass flow tank based on the design used by Dr. Steven Vogel, of Duke University. Steven was a real supporter of our work. In fact, if he hadn't assured me that we could use water to accurately simulate airflow, by adjusting flow rate for kinematic viscosity, this work might never have been done.

Science Survival Tip No. 13: *Do not reinvent the wheel if you can copy someone else's design, but remember to give them credit.*

Giving credit to your students is also critical

I was invited to speak at a meeting in Washington, D.C., to present work on the mathematical simulation of glycogen regulation in the liver [133]. Why glycogen, you may ask? Well!

War is hard on kids, especially when it comes to the effect of diet on their longterm health.

As a child of war, raised on the ruins of an English port city, I was almost always hungry. Our mother did well, harvesting edible wild plants and choosing the least expensive, but most nutritious, options. I still remember the smell of neck-of-mutton soup, simmering on that old gas stove in the light of the glass covered kitchen. With five teenagers, there never was enough food.

Rationing books and the black market reigned supreme. If you had money, you could buy milk, butter, vitamins, almost anything. If not, you were confined to the small portions permitted by the family's rationing book.

The continual desire for food made my favorite place the school cafeteria. There I could eat five solid lunches a week, often augmented by the richer kids leftovers. Hunger leaves an indelible impression on the mind. It can lead a person to hoard food later in life, to purchase far too much at the store, and to become an avid vegetable gardener (which I still am), fisherman or hunter.

I also found my work tended to focus on the role of feeding in cellular biochemistry.

A young Chinese scientist, Ke, and I had been working for almost two years on the regulation of liver glycogen. If you do not eat, liver glycogen eventually becomes depleted. After you eat a meal, the carbohydrates taken in are either burned, stored as glycogen in the liver, or converted to fat. We were interested in understanding the regulation of these events at the cellular level, using a mathematical model combined with some of our transcriptomics data [124].

I had a 45-minute speaker slot at a Washington meeting. Ke knew her math better than me by far, so I offered to share the podium with her. I addressed the biochemistry aspects of our research, while Ke presented the applied math components [133]. Ke was delighted, did a great job, and was offered a position as a direct result of someone in the audience hearing her presentation. These are the kinds of events that grease the skids of science, while rarely if ever being mentioned in scientific articles.

Back to 1991 and Steven Vogel's Lab

Having been invited to present our work on nasal airflow at Duke University, my artist friend and collaborator on this project, Andy Fleishman, and I decided to talk about the importance of the interface between art with science. We were building models (molds made from casts) of the nose of the rat, using water to study airflow. Air turned out to be an impossible medium because of the tiny size of the airways in our nasal molds. I stated this concern during the talk. At this point, a member of the audience, Steven Vogel, an expert in fluid flow at Duke, politely allayed my concerns.

Earlier, a member of our scientific advisory panel, Dr. Jones, had expressed similar reservations about the use of water in place of air. I had attempted to see a dense smoke flowing through the mold, and it was completely invisible in the narrow mold passageways. He insisted at our annual review that it was possible to study gas flow in these molds. The following year, Dr. Jones repeated this opinion, and insisted that we stop using water and study airflow directly.

Scientific advisory panels can negatively influence your funding. They are dangerous. Having read *The Art of War*, by Sun Tzu, I was ready for his attack on our approach. I reached into my pocket, in front of the advisory panel, scientific staff and management of our institute, and pulled out a perfect rat nasal mold. It was plumbed up and ready to use. I walked over to Dr. Jones, handed him the mold and said, "Excellent, maybe you can work it out, and tell us how." Dr. Jones accepted the gift, somewhat reluctantly, and I forgot about it.

Six months later, a 6 inches square parcel turned up on my desk. It contained the mold I had handed to Dr. Jones, along with a terse handwritten note stating, *It can't be done!* I wisely decided to not say a word about this incident the following year, and nor did Dr. Jones.

Science Survival Tip No. 14: *Do not publicly say "I told you so," it achieves nothing useful.*

When Steven Vogel invited our group to visit his lab. There, he had the flow tank that we had copied for our studies of boundary layer effects on X-ray film development patterns. This tank was designed for studies of squid swimming mechanics. He also had a large wind tunnel in which he demonstrated the flight of a Harris hawk. The hawk was placed on a perch in a clear observation chamber. The bird was calm and seemed to be content.

Steven started the wind tunnel, which made quite a racket. As the air speed slowly increased, the bird's feathers ruffled, then she slowly extended her wings and gently lifted off. This lovely hawk was flying a foot or so from our noses. A truly remarkable experience for all of us. This is nowhere to be seen in any of our articles, but it sure was important for our continuing studies of nasal airflow.

That is how science works. Inspiration from people and events never mentioned in publications.

―――――

Back to Jim and our squid-tank lookalike. This flow tank was 4 feet long, 3 feet high, and 1 foot deep. It had a motor and propeller, to circulate water along the top, down to a separate chamber along the base, and back along the bottom, to return to the upper chamber via a flow straightener made of tightly-packed, plastic drinking straws. It was essentially a toroidal tube, with a square cross-section, designed to create laminar flow, with access from above.

Jim immediately understood what we were trying to do. He knew that our interest lay in understanding gas deposition in the nose, but he did not ask the question that used to irritate me no end, "Why isn't your flow tank shaped like a nose?" When I tried to explain, most people's eyes would glaze and they would wander off, surely thinking, "Morgan's weird."

Jim came up with a bunch of ideas that we had not yet tried. Each of which was tested and discarded.

Then came his phone call.

So many blind alleys and dead ends that had spawned successful careers for several young scientists. This work became more and more mathematically based [71], with our beloved flow tank generally standing idle.

That said, reading those papers in the appendix gives you no idea about the nature of our day-to-day work. All the head scratching and equipment failures are never mentioned. Publications explain the end result, which could mislead people considering a career in science. Its more fun and trickier than the publications would lead you to believe.

Scientific articles do not mention our working in the lab until midnight with airway casts, festooned with wires attached to an array of ammeters, or messy multicolored soggy paper, extracted from the tank, or clusters of particles on the floor of the tank, which told us nothing about potential gas uptake rates in the nose. And, of course, the enthusiastic, and sometimes heated, discussions over a beer or a meal after work.

The real work of science, repeated failure, is never mentioned. If you do put them in the narrative, the editors will strike them out, or worse, reject your article completely. I often wonder if those people had actually done any research themselves.

Science Survival Tip No. 15: *If I had an hour to solve a problem, I'd spend 55 minutes thinking about the problem, and five minutes thinking about solutions.*

— Albert Einstein

Where did Jim's solution come from?

The solution seemed to come in the brief five minutes, as I cranked up the flow tank full of dilute developer, and a remarkably precise pattern emerged. I was looking at the structure of a complex boundary layer. But do not think that result came in a flash.

Where did it come from?

- It came from Dr. Steven Vogel encouraging our continued studies with water flow as a substitute for air.
- It came from whoever helped Dr. Vogel design his squid flow tank.

- It came from Bobby, the person who built that tank.

Bobby was a skilled technician and engineer who created many small pieces of lab equipment for our team. He only took one look at Dr. Vogel's tank and recreated it from scratch. Bobby's engineering never let us down with leaks or motor failures.

Bobby had a drinking problem, which was no secret. Bobby's wife always wanted him to learn to dance with her, and he always refused. Bobby took one drink too many, and went into liver failure. Fortunately, rather than dying right then, he received a liver transplant. Bobby told me later, that while lying in the intensive care unit with his wife at his bedside, he promised that if he survived he would take dancing lessons and take her dancing.

Bobby did both of these things, and lived for many more years to dance with his wife instead of drinking.

There is a valuable lesson in there, and I'm grateful to Bobby for telling me his story.

- Jim's successful idea also came from all those failed attempts, with mucus, biochemistry and blood flow.
- It came from my original observation of the lesion in the nose.
- It came from the original experiment on formaldehyde carcinogenicity, published in the 1970s.

Science Survival Tip No. 16: *It's OK to stand on the shoulders of giants whenever you get a chance. The view is better from up there.*

By the way, we all need heroes to guide and inspire us. One of my heroes is Jean-Dominique Bauby. If you think you're having a bad day, read his book, *The Diving Bell and the Butterfly*, and think again.

PRIDE VERSUS PLAY

Science Survival Tip No. 17: *Any effort that has self glorification as its final endpoint is bound to end in disaster.*

— Robert M. Pirsig

I learned a lot working as a country veterinarian on the farms of England. Don't be too proud was one important lesson.

1970 - Pride Cometh Before A Fall

The last bovine caesarian section I performed before leaving veterinary practice for a research career in 1970 was during the final weeks of working for a great veterinary practice in Wellington, Somerset. A lovely village that would remind you of those wonderful veterinary stories by James Herriot, in *All Creatures Great And Small*.

I was called to assist a cow in labor in the middle of a rainy afternoon and when I say rainy, I mean a continuous downpour. A burly looking farmer with a deep voice met me at the farm gate, and it was clear that he considered himself to be in charge. He did not look particularly pleased to see me. I was a skilled veterinarian by that time, but I am only 5 feet 6 inches tall, and I looked young. Some farmers would complain that the practice was sending children to treat their stock. This guy was a bit like that.

He marched me to a dairy cow, standing in a nearby paddock. She was noticeably ready to deliver her calf. I checked inside the womb and there was no way to turn this calf, which was kicking and very much alive. It was at an angle that I just could not fix in the routine manner.

I had successfully delivered many difficult calves with calving ropes and a little patience during three years in practice, but not this one. "I'm afraid she'll need a caesarian," I said, but it won't take long. This farmer clearly loved his cows, as he didn't hesitate to spend the money on the operation.

I extracted my gear from the car, and sterilized the instruments in a little boiler designed for the job that I set up in the farmhouse kitchen. The instrument tray was placed on a clean towel covering a bale of straw next to the expectant mother. She was a large, good-tempered Holstein-Friesian.

It continued to downpour. I handed the truculent, but likable, farmer a large umbrella, and asked him to hold it over the operation site so I could see what I was doing. He seemed to think that this job was below his station, but he complied. Out of pure curiosity, I am sure.

I injected a tranquilizing drug, to keep her calm, followed by a paravertebral nerve block. The mother to be, though soaking wet, was all ready for surgery after I had shaved and sterilized her left flank. I proceeded to make that first incision through the skin, a cut about 15 inches long. It takes a large opening to deliver a 100-pound calf.

Then, something off to my right caught my attention. The farmer, still holding the umbrella in a way that reminded me of Mary Poppins, was falling like a felled-tree. His body crashed directly onto the instruments, and into the mud went everything, farmer and all.

Once I was sure the cow was safe, I promptly checked the farmer was in no danger, and dragged his inert body to lean against an adjacent tree. I remember James Herriot saying in one of his stories that it is always the big guys who faint at the sight of blood. Well, here was another one. I hurried to find the farmer's wife, a tiny woman with lively eyes and a quick whit. On hearing the news, she told me that her husband was a big milksop, and she was not surprised at the mess he'd made.

Within another 20 minutes, we had sterilized the instruments, and with the lady of the house holding the umbrella, we delivered a lovely healthy heifer calf. And, even after the delivery, it was still raining! The farmer recovered from his fall, his only injury being dented pride, and we all dried off and enjoyed a cup of tea in their cosy farmhouse.

1974 - Dr. M's Pride Spoils His Visit

I had just published an article on a brain disease of sheep called polioencephalomalacia [2].

- Polio = grey.
- Encephalo = in the skull meaning the brain.
- Malacia = softening.

This article drew conclusions contrary to those of an "expert in the field, Dr. M, a man I respected based on his published works, but had never met in person.

I was pleased with this article. It was based on the results of findings I had made using an electron microscope to study the brains of three dying sheep. The article referred to the work of Dr. M, who was well known for his studies of diseases in sheep.

A few months after the article was published, Dr. M wrote to me, to say that he would like to visit to discuss my conclusions. He did not say findings, he said conclusions. I was young, so this went right over my head. This was the first report of ultrastructural changes in the brain in cases of ovine polioencephalomalacia.

Excited and pleased about the meeting, I looked forward with anticipation to learning more about maladies of sheep. Dr. M had 40 years of experience in this field, while I had only three. The man made a 400-mile train journey, followed by a lengthy bus trip, to reach our small sheep diseases research institute. He was not what I had been expecting.

Dr. M clearly planned to intimidate me, but he had never encountered the mother who raised me. When he arrived, I was working in my office, surrounded by books, a manual ultratome, dozens of photocopied manuscripts, my beloved research microscope, boxes of glass slides, and other pathology paraphernalia. There was a knock on the door and in barged Dr. M.

He was a fairly large man, in his late 60s, as I remember, and he was soaking wet from the incessant rain that day. He did not respond to my friendly hello. On hanging up his coat and hat, he sat without preamble, and stared at me, for what seemed like ages.

My offer of a nice hot cup of tea received no response. The man was angry. Angry about my publication. His first statement was one long sentence, containing a four-pronged attack on my work and me personally. I remember admiring his sentence structure:

1. *What gives you the audacity* [talking down to me, as he had a Ph.D., and I didn't, yet],
2. *to claim* [a put down designed to intimidate]
3. *that the primary biochemical lesion of* **cerebro-cortical necrosis** [his emphasis, I had used the more appropriate North American name for this disease, *polioencephalomalacia*]
4. *lies in the astrocyte?* [contradicting his conclusion that the primary disease events occur in neurons].

Great way to start a conversation with a neophyte researcher, who is all excited about his work. I was initially shocked and affronted, then pissed, before I calmly stepped into logic mode. This is a powerful tool. Almost as powerful as silence, which Dr. M had used to good effect.

I am one of those fortunate people who goes silent when angry, just in case I say something I would regret. Where I learned this, I have no idea, but it is generally the better course of action. If I am really pissed, I go for a long walk as gradually reason returns, and then I address the issue as best I can.

Science Survival Tip No. 18: *In science, it does not matter who is right, it only matters what is right.*

I attempted, without success, to debate our difference of opinion on the basis of the way the brain lesions develop and the biochemical physiology of astrocytes versus nerve cells. I do not remember his providing any kind of logical rebuttal.

Science Survival Tip No. 19: *In conflict, be fair and generous.*

— Lao-tzu

After a few minutes of my futile attempts to trigger a meaningful conversation, he stood, put on his wet coat and hat, and left my office, without a backward glance. This goodbye disappointed me, but it provided an important lesson to never ever treat young scientists in this manner.

Science Survival Tip No. 20: *Be good to people on the way up, you may meet them again on the way down.*

Sadly, Dr. M died a couple of years later, and I wondered if he was not himself during his weird visit. Maybe he was just sick? People warned me that he could be prickly. That was not prickly, it is not what science is all about.

Prepare to be attacked if you challenge someone's precious ideas, especially if you do so in public. Don't be like Gollum, and make your ideas *your precious*.

That said, you had better fully understand what you are talking about if you plan to challenge other's ideas, or do not go there. This is one of my favorite aspects of science because it forces me to step outside of my comfort zone and learn. If you are not attached to your ideas, and you consider other people's opinions carefully with respect, you will be fine.

1990 - The Letter

I was sitting in my office one day, when Julie came into my office and said, "Kevin, look at this letter. What do you think of Fred's edit?" Fred was her boss. She was not sure if she should make the change or not. It was an important letter, about a publication or a grant, I forget which.

Halfway down the letter, Julie had written, "I'll need to play with the equations."

Fred had crossed out play, and substituted, work.

Fred was wrong. Julie wisely left it as it was.

Creativity comes from play, not work. If in doubt, watch how young animals learn to master the skills they will need to survive as adults.

They play.

Science Survival Tip No. 21: *Do not let stodgy people put down your inner child, to turn creative play into tedious work.*

Winter 1959: Black Snow

I was a 17-year old boy, home alone in Bristol, England. I was truly alone, as the rest of the family had moved away, and Mum had given me the keys of an old, empty dusty house, that had not been lived in for years. She made a living buying almost derelict properties and fixing them up for sale. I lived in a sparsely furnished, ground-level room, in the front, in an otherwise empty, dusty two-story, three bedroom building.

I was bored. Dangerous combination, bored and teenager.

I had enjoyed life in this house for about six months while going to school during the week, and working in a hotel on weekends to earn money for food, school supplies and utilities, including coal for the fire. I was alone and in peace, while completing the sixth form (senior year) at the local grammar school.

My one furnished room had a single bed, an old couch, a wooden table, a few clothes, my water polo and motorbike gear, a school bag and a fireplace. It was in the middle of winter, dark outside and late in the evening with homework done. There wasn't much to entertain me apart from reading or sleeping. I cast around for something to do.

Then I remembered the bottles of used methanol previously employed by my biology teacher for storing biological specimens (e.g. dead frogs). He had given me two 1-liter bottles of this waste, slimy green preservative, as a gift. I was known to be an avid biologist, and he wanted to get rid of them.

I wonder what methanol is like when it burns? I know it's flammable.

With a good coal fire burning in the grate, I decided to throw a spoonful of the green liquid onto the fire. What a lovely surprise! A bright reddish-yellow fireball filled the grate for a fraction of a second and then was as quickly extinguished. This was followed by a popping sound. I enjoyed that, and the temporary wave of heat felt good, too.

Why not try a small cup of methanol? This was becoming exciting. It yielded a more impressive result, a mass of flames, engulfing the entire mantelpiece. There was a beautiful dark blue core, surrounded by bright yellow tendrils of burning methanol, resembling living snakes reaching out toward me. Then, once again, a pop followed by an odd silence.

Maybe it's time to be a little cautious.

Caution for a teenager, with my underdeveloped frontal cortex, involved turning the couch around to create a barrier between myself and the next experiment. Hiding behind the couch for protection, and with a little (but not enough) trepidation, I lobbed a huge stream of the flammable liquid onto the fire.

Wooosh!

A massive fireball engulfed the entire wall, extending to the ceiling, while lapping around the margins of my protective couch. Again, the flame was as quickly extinguished, followed by a loud pop, leaving an uncanny silence. I enjoyed that, and was considering a second bottle, yes, I had two, when I noticed a black smudge on my hand. Then another!

That's odd.

Glancing upward, I was horrified to see a swirling mass of black snow-like flakes above my head, filling the entire top of my living quarters. These flakes were gently descending upon me and my few belongings. The black snow was evidently pure carbon (C), released from the methanol (CH_3OH) as a fine black powder. Incomplete combustion as a consequence of a limited oxygen supply in the center of the inferno. This yielded fine carbon dust, rather than carbon monoxide and carbon dioxide gases. The precipitated carbon had taken a form that resembled black snowflakes. They were beautiful, and they were descending inexorably onto everything I owned as I was becoming covered with black smudges.

Having experienced the joy of incomplete combustion, I grabbed my school bag and a clean shirt, leaving my homework open on the table, and crawled along the floor to the door. Everything left behind was soon coated with a layer of fine black carbon. I explained this to my teachers the next day, as having been because of a fire at home, which was technically correct.

Damn, that was fun. I would not have missed it for anything.

Then there was that sheep, many years later.

Fall 1969: Setting the Barn on Fire

A call came into the practice, from a remote farm.

A bloated sheep.

You have to get there quick, or they will die. I took the job off the book, and rushed off in my Morris Traveller. They have plenty of room for calving ropes, surgical gear, and an array of medicines and anesthetics.

The ewe, pronounced *yaw* in Somerset, a female sheep, was lying on dry straw, in an old barn. She was struggling to breath, gasping for air. The cobwebbed barn was full to the rafters with straw bales. There was a clear area by the door under a window for us to do our work.

The farmer, approximately my own age (late 20s), was a nice chap. During my examination, he asked me about bloat. It is often caused by a change of diet, such as going from grass to rich clover. Methane gas accumulates in the rumen, the largest compartment of the stomach of sheep and cattle. This results in abdominal swelling, pressure on the chest, and asphyxiation. Unless the rumen gas is released, which is where I come in!

I extracted a trocar and cannula from the car. The trocar is a sharp, metal spike, about 6 inches long, designed to penetrate the tough skin of sheep and cattle. It fits snuggly inside the cannula, a hollow tube with a flange, which is used to stitch it to the skin if need be. The trocar-cannula is inserted just behind the last rib, high on the flank, near the spine, into the gas pocket, in the rumen. After removing the trocar from the cannula, the gas escapes, to the relief of the bloated animal.

I had the unit installed and was about to withdraw the trocar from the cannula. Interrupting me, the farmer, said, "Kevin, I heard that this gas is flammable. It burns! Is that true?"

I replied that it was. He then asked me if I had ever ignited the gas. I had not, but I was only too happy to give it a try, if he wanted me to.

Boys love to play with fire.

He immediately reached into his pocket, and pulled out a cigarette lighter, as it had been planned all along. I had considered the sheep to be at no risk. I pulled out the trocar. Gas escaped with a hiss.

He reached out with the lighter.

WOOOOOSSSHHH!

A bluish-yellow blaze, about 8 feet long, shot out of the sheep's side. Just missing us, it set light to the pile of straw bales. The fire spread and we came close to burning down his barn. After beating out the fire with old sacks, we fell about in hysterical laughter.

The sheep was fine, by the way. I was a conscientious veterinarian.

I bet that story about the 'young vit-n-ry and a flaming yaw' was worth a pint or two down the local pub that night, and for years to come.

I never did dare to try it in a cow.

KNOW YOURSELF

Science Survival Tip No. 22: *Find out who you are and do it on purpose.*

— Dolly Parton

My childhood, like many children of war, was marred by family conflict. I attempted to not exist during these storms of emotion and broken crockery.

My first fond memories are of animals and plants, especially those living in freshwater ponds. The more green slime the better. Green is the color of life.

My approach to the emotional turmoil of my home was to go inward, to close down emotions. It was much safer that way. I became introspective, finding joy in a series of hobbies. I became more attracted to thinking than expressing feelings, which tended to set me apart from my peers as they approached their teens. Rather than the pop music and the dancing of my friends, I preferred quiet reading and study.

Science Survival Tips No. 23: *Cut those unhealthy childhood tapes in your head, for real freedom of thought and personal happiness.*

The seasons came and went. Frogs eggs turned into tadpoles, and tadpoles into frogs. Many years later, I learned much about the making of frogs during studies of embryology. A burning passion to understand the nature of life encouraged me to read widely. In fact, it was reading that allowed me to go to a university.

As much as I loved nature, I had problems. Difficult emotional family circumstances resulted in me becoming a slow learner, especially when it came to two critical subjects, reading and arithmetic. If you didn't pass the 11-plus exam, in England, 1954, there was no way you would ever go to a grammar school. The road would then be closed to a university education. I had to pass the 11 plus, but I had no idea that was the case.

I took it at 11 years old and failed, but my elder brother and sister had passed and gone into higher education. I felt bad about myself, certain I was really dumb, like Mum had said. Self-doubt because of negative parenting can result in a terrible self-image. Here are some direct quotes, that I heard a thousand times, even in my 40s: *You're short, and your eyesight's no good. You can't find your way anywhere. You have a terrible memory, while mine is perfect.*

And my favorite, *You're just like your Aunt Alice, she couldn't do anything right, either.*

I do not think this was really about me, rather than Mum feeling bad about herself. She regretted not having a college education while being well read. War is tough on people. Otherwise,

Mum did a great job keeping us alive and fed, and finally encouraging our education even at considerable cost to herself. I was mad at her for many years, but now, having raised kids myself under much easier circumstances, I appreciate the remarkable things she achieved, especially when it came to encouraging us to read, learn a foreign language and get a college education.

At 12 years old, a friend told me about a great movie he had just watched at a movie theatre on the other side of Bristol, two bus rides away. We were free to do as we pleased, and having earned a few shillings by helping a friend's dad untangle miles of binder twine, I planned my adventure one Saturday morning. Funny how I can remember it so clearly, over half a century later.

What a movie. A musical about the tales of Hans Christian Anderson. It included a wonderful story about an ugly duckling that turned out to be a swan. It had happened just at the right time. I realized during that movie that I was being described as an ugly duckling by my mother, and I was saved from a negative self-image by Danny Kaye and the power of storytelling.

This played a critical role in my success as a scientist years later. Science can beat you down, and a negative self-image can kill your chances of success, but I was always a scientist at heart.

Can you imagine me, as a 10 year old, carrying a jam-jar with a string handle fashioned by my Mum, to collect frog's spawn and other treasures for our fish tank at home. Yes, Mum helped us make and stock the tank. She was a good, but troubled person.

I would lay for hours on the bank of a stream or pond. Keeping still, so as to avoid creating a disturbance, while admiring the comings and goings of ducks and other birds. I would even spot hedgehogs and foxes from time to time. On looking down, I especially liked to watch the sticklebacks making their nests.

I marveled at the way these shiny little fish would busily stir the water with their fins to create a depression on the sandy floor. Then they would take plant debris in their mouths and arrange a protective home for their hatchlings. This pond was a safe place for me, too.

I found great joy in a series of hobbies, including water polo, photography, and botany. I spent countless hours trying to distinguish petaloid sepals from sepaloid petals, with my copy of *a Flora Of The British Isles* under my arm, as I explored the biodiverse hedgerows of England. This was great for intellectual pursuits, but useless for developing my social skills.

A Prestigious Award

In the fifth form at grammar school, at fifteen years old, I received a prize of 10 pounds, an enormous sum. It was for achieving one of the highest grades at the end of term exams, in some subject or other. What do you think that 15-year-old kid spent it on? I could have bought anything I wanted, and I did.

I chose a book on the classification of British grasses. On proudly displaying my purchase to my classmates, they went strangely quiet. If I had wasted the money on football or cricket memorabilia, I would have been regarded as a cool kid. People really are weird, and by some stroke of luck, I am normal.

Many years later, as a successful scientist, nothing had changed. I was sitting in the cubical of one of our post-doctoral students, sorting out a statistical issue. I was quietly enjoying myself, hidden away, when Michelle, a fellow scientist, sticks her head around the entrance to the cubical, and waves a copy of *People Magazine* in my face, saying,

"Kevin, I've finally worked out what's wrong with you. You have nine of the ten characteristics in this test."

"What test?" I reply, coming out of my analytical trance.

"The test for eccentricity. You are eccentric. I often wondered what was wrong with you."

It is fortunate I love being me!

How I Became a Veterinarian

I was sitting in front of the careers master, at St. George's Grammar School, in Bristol, England, 1960.

"Kevin, we have discussed your abilities in the staff room, and we think that you would make an excellent biochemist."

"I was thinking of going to medical, dental or veterinary school, sir." I replied.

This decision was based on a crazy desire to play clarinet in a dance band and travel the world. I was completely unsuited for such a dream, and I was not a great clarinet player, but the careers master made a reply that would define my life:

"Well, Kevin, I am sure you could get into the medical or dental schools, but no pupil from this school has yet to enter a veterinary school."

The English, for all their strange ways, love their animals. Not prone to expressions of emotion, they will bill and coo over dogs, cats or budgerigars, almost with tears in their eyes.

It was decided, "Well sir, let's see if I can get in!"

And I became a vet, but it did not turn out to be my element.

February, 1969. 3:00 am.

The phone is ringing, again.

I have been working for three days, nonstop. Driving from farm to farm, day and night, all over a 25-mile radius from our veterinary practice office, in Wellington, Somerset, England. I am all in, and not happy. Tension is building. I am wondering if I'm cut out to be a veterinarian in clinical practice, even though most clients seem to appreciate my work.

Where did that thought come from?

The unhappiness was slowly growing. I had been in this line of work for three years following graduation from the Bristol Veterinary School. Three different jobs. This one is great, with only one night in six and one weekend in six on call. My fellow veterinary colleagues were a collegial group of guys, and the animals are fine, but you need excellent people skills in the veterinary profession. Not my strong suit.

It has been a long weekend on duty, night and day, and the

phone is ringing at 3 in the morning. I get out of bed, go downstairs, and stand next to the phone in the cold hallway. With trepidation, I answer the call.

I hope it's not a difficult calving, 20 miles away. I'm exhausted.

An elderly man's voice says, "Is that the vet?" I grunt my sleepy reply, and he says, "It's about my cat." I have no desire to leave the house again, especially on a cold night, but my job is to assist animals and their owners. I start the usual line of questioning about symptoms. Then I ask the question that ruined the rest of the day, while freeing me to choose a whole new existence, better suited to my personality.

"How long has this been going on?"

"About a week."

"And you call me at three in the morning?" I nearly scream.

I was immediately angry. There was no urgency. It could wait until morning.

He reacted with "Well, it's your job, and you're on duty, aren't you?"

I explained that we are a small practice. There are no eight-hour shifts. After-hours calls are for emergencies.

Some people are inconsiderate. At that age, my psyche was ill-equipped to cope. I reassured the man that it could wait until morning surgery. He was appeased and agreed to be there. I did manage to apologize for my frustration and calm my voice, but not my anger.

Then I felt it. Between my shoulder blades. A massive muscle spasm that confined me to my bed in intense pain, unable to move, for the rest of the day. I was just standing there, talking. It would appear that just standing and talking can be perilous, if you don't have your mind under control. What was the real message my body was sending me, you might wonder?

While lying in bed, barely able to move, I realized that I was not in my element. I needed to seek other work, which I did. Six months later, I entered a career in science, as a pathologist and then a researcher. I never looked back. My temperament is better suited to solving scientific problems, with studies that span years. Veterinary practice deals with issues that generally take minutes. It also requires great people skills, while diplomacy has never been my strength.

This was not about a sick cat.
This was about a veterinarian's life out of whack!

Science Survival Tip No. 24: *Listen to your body, because it will let you know if you're in your optimal job.*

FIND YOUR ELEMENT

Science Survival Tip No. 25: *For most of us the problem isn't that we aim too high and fail - it's just the opposite - we aim too low and succeed.*

— Ken Robinson

Unhappy as a country vet, and to cut a long story short, I applied for about a hundred different positions to finally be offered a job in Edinburgh, Scotland. I was hired as a comparative neuropathologist. I didn't know anything about it at the time, but I was essentially the only candidate. The pay was poor, the climate cold, the Scots suspicious of the English, about which I'm not surprised.

Having relocated the family to this cold, windy country of mists and rain, golf and whisky, I entered the Moredun Institute, a cluster of old brick buildings and sheep pens that were to become my workplace for the next five enjoyable years. I filled out some paperwork in the administrative office, and was directed to the pathology department.

No other instructions were provided, apart from "Dr. Barlow said you can use his office until he returns from summer holiday." This was delivered by a terse Scottish histotechnologist. I could barely understand her thick brogue.

Into Dr. Barlow's unfamiliar, small dusty office, I went. The first thing I noticed, apart from the many sheep outside the large window, was a jar of tadpoles on a window ledge.

What are the tadpoles about? I asked. The young lady, in a multicolored, stained lab coat, replied, "Dr. Barlow is trying to create hairy, shaking frogs."

Um, interesting.

She left without further ado, leaving me in a perplexed silence. Then the truth dawned on me. *Three paid weeks in this office. No orders to follow. No cows or cats to fix. No farmers to placate. Do whatever I want. Come and go as I please.*

As soon as I recovered from disbelief, I was ecstatic.

I've come home!

The small office was full of books on neuroscience and sheep diseases, piles of scientific articles, shelves covered with sheep skulls and other fascinating biological samples, boxes of microscope slides and a fancy microscope. I scanned some of the textbooks with interest, and then sat at the scope, with some trepidation. I sure didn't want to break it on my first day. It was a complex research scope, with lots of dials and nobs, that I would come to master in due course.

Taking a slide box out at random, I extracted a 3-by-1 inch glass slide, put it on the scope, and was immediately entranced. It contained a section of the brainstem of a sheep, stained with Luxol Fast Blue for myelin, to show large nerve cell processes. The plaited network of myelinated axons of the reticular formation was beautiful, and it actually took my breath away. A far cry from a sick cat, as much as I love cats.

Eventually, Dr. Barlow returned and put me to work.

1970: My Primary Trade Becomes Pathology

Pathology: The study of the essential nature of diseases and especially of the structural and functional changes produced by them.

— Merriam Webster Dictionary

Accepting the position at the Moredun Sheep Diseases

Research Institute settled my academic fate: I was to become a pathologist. This was not something I had dreamed of doing, planned for, or expected. It was the result of my escape from veterinary practice, and having to deal with those capricious higher primates, the public.

My boss, Dick, was a great guy who trained me in the art of cutting up sheep and cattle brains for microscopic examination and diagnosis. The main conditions I saw were scrapie (prion disease), looping ill (viral infection), focal symmetrical encephalomalacia (bacterial toxin), polioencephalomalacia (bacterial thiamine antimetabolite poisoning), listeriosis (bacterial infection) and copper toxicity (metal poisoning), with a scattering of rarer conditions such as storage diseases and grass sickness (horses).

I then sent out a one-page report to the person who provided the specimen. Most of the time, these were veterinarians working in the field service. They were called in by farm veterinarians who ran into diagnostic problems. I did this for all of Scotland, usually a few hundred samples per year. This left me with plenty of time to think of other things to do, which is how I became a research pathologist.

People brought me interesting cases, as most general pathologists tended to be a little squeamish of diseases of the nervous system. That was no problem for me, as that was all I studied all the time. I hadn't chosen neuropathology, it chose me.

"It is our choices, Harry, that show what we truly are, far more than our abilities."

— JK Rowling

Choose your trade with care

Many years ago, I was an enthusiastic flautist in my mid-20s. Though not a great flute player, I regularly performed in local amateur chamber groups and the town orchestra. One late fall day, in Penicuik, Scotland, I was pleased and terrified to be invited by a local dramatics society to join the small pit orchestra, for a performance of *The Pirates of Penzance* by Gilbert and Sullivan.

The process was simple. You are handed your part on Sunday morning, it having been released from a previous performance elsewhere. You have the morning to practice the tricky bits, then you turn up for the orchestra rehearsal later that afternoon.

I checked the score for pit-orchestra, designed for nine instruments, including one flute. It looked challenging but doable. There was one part that made me nervous. It was fast, with 11 repeats of an entire page of music. This means you have to keep track, while playing, of how many times you have played that page.

At the first rehearsal, I found myself sitting next to an excellent clarinetist, Tom. I had given up the clarinet due to the pain of changing reeds and the fact that this instrument over-blows a 12th. Octaves are easier.

Tom was much older than me, being in his early 60s. A nice man, a little portly, with a fringe of greying hair and a delightful smile. He produced a wonderful tone with his instrument. During this first rehearsal, Tom accidentally kicked over his music stand. Multiple pages of loosely bound sheet music tumbled to the floor.

If I had done that, I would be doomed. I was pretty good at sight reading, but could never remember more than a few notes of any piece. I was basically a "technical flautist." Tom kept right on playing, as if nothing had happened. In fact, he didn't pick up his music from the floor until we came to a break. He was playing the whole thing beautifully from memory.

During the break, I told Tom how impressed I was with his performance, and asked him if he had considered playing professionally. Tom replied that he had played semiprofessionally, but the competition was so fierce that he was forced to abandon his dream and work as an electrician. *Now I just play for fun,* Tom said, with a wistful expression tinged with sadness. He'd had a dream that he was forced to abandon.

I have seen similar experiences in science, which is equally competitive, especially if you want to do basic research. I was fortunate to guide a young scientist away from a potential tragedy in this respect.

About 20 years later, in the mid-90s, I was working as the postdoctoral coordinator for a small research foundation in which I had an active laboratory. In addition to guiding my own graduate students and post-doctoral fellows, I provided support to about 40 other young scientists. My job was to answer their questions and negotiate disagreements with the institute administration, and in rare events, their mentors.

One particular post-doctoral fellow had a large, framed photo of herself in her cubicle. She was holding her doctorate, rolled up in a scroll, and looking proud of her achievement. This image gave me the willies. It was clear that this person was attached to the image of science, while I knew her strengths lay elsewhere. Her interpersonal skills outstripped mine by miles. I thought she should be a tour guide rather than a scientist.

This lady came to my office one day to enquire whether I could provide her with a letter of reference for a job in academia. She was applying for a research position in a prestigious (read competitive as hell) laboratory in an Ivy League university. It would have been disastrous!

I assured her that I could write a letter of reference, but I thought her choice was misguided. In fact, my lack of diplomacy resulted in tears that I had not intended. After a long conversation, she applied for a study director position with an industrial company. A study director job is not so dissimilar to a tour guide. They plan ahead, keep track of lots of little details, work to be sure no one gets lost, and when the task is completed to everyone's satisfaction and on time, there are smiles all around.

I was pleased that this delightful person heeded my advice, and many years later I heard that she had excelled in her work and was much sought after as a study director.

Think about it before you start going down the wrong road. Not everyone is cut out for research. We also need excellent managers, study directors, technologists, statisticians, tour directors and so forth.

Find your element, and you'll almost certainly do well.

I Slowly Master My Trade

Pathology is a much needed skill in many fields of biological research. That is what you need, first and foremost: A marketable trade. You now become part of a team, working with scientists in other disciplines. To be effective, you have to learn how to speak their language. Not as well as them, but well enough. This was what I did. Not as a plan, just in response to wanting to understand diseases that came my way.

If there was a biochemistry problem, I would study biochemistry. I did this by reading textbooks for fun. My favorite biochemistry book was *Harper's Biochemistry*. Written for medical professionals and regularly updated. In my career, I watched that book evolve over a period of many years. If it was an immunology issue, I would study that, and so on. You can not afford to stand still. The world changes, so change with it.

Science Survival Tip No. 26: *Be unreasonable - The reasonable man [person] adapts himself to the world; the unreasonable one persists in trying to adapt the world to himself. Therefore, all progress depends on the unreasonable man [person].*

— George Bernard Shaw

I proceeded to learn as much as I could about every other discipline I needed to solve the pathology problems that came my way. This involves plenty of learning, which usually starts with a textbook. I highly recommend reading them like novels, as they usually tell the story of a person's life-long study of a particular subject.

I would then start to dig deeper in relation to the particular problem I was attempting to solve, while seeking experts willing to advise me, or better still collaborate in the research. This approach tended to allay major screwups. In exchange for their help, I would offer something in return, such as a co-authorship or assistance with interpreting some tissue slides. You can even trade use of your lab, technicians (with their compliance), or ideas.

That said, my first scientific articles were sole author. This is rare, especially today. Once I had a couple under my belt, I realized that the senior staff around me either had Ph.D. after their name, or they were working on it. I asked Dick if I could undertake studies for a doctoral degree. He replied, "That's fine, Kevin, as long as it doesn't interfere with your work."

A PhD Is Not Enough, But It Helps

Today, in the United States, education has become a business, and all of my graduate students had to take excessive course work. Graduate course work, mandated by Edinburgh University in the 1970s, was one course of the student's choosing. I was even allowed to simply audit a course. Why would I need an exam? I was doing it to learn, not for a grade. In the end, I chose to audit master's level biochemistry. One year of free biochemistry lectures.

Fascinating!

While doing the research for this book, I came across *A PhD Is Not Enough* by Peter J. Fiebelman. The narrative contains many gems, but I was struck by the author's journey into science. It was so different to my own. Dr. Fiebelman wanted to do research from an early age, so he followed the arduous path of academia. My goal was to be gainfully employed while enjoying life, with no idea how I intended to do that.

I succeeded in research by chance, gaining a back-door access to a research career. The back door of a much-needed trade.

This door is still open today.

There are many roads to a successful and enjoyable career in science. That said, something Dr. Fiebelman and I had in common was the good fortune of encountering remarkable mentors.

A Childhood Hobby Carries Me Into Research

I spent much of my time looking down a microscope, for which I was well prepared. Let's go back in time, to find out why.

1956

One rainy day in downtown Bristol at 13 years old, I was seeking yet another book by Edgar Rice Boroughs (I loved his Mars series). I can see it now. A dusty second-hand bookshop, run by an extremely shortsighted, heavy-set man. He would put his thick lenses within an inch or two of anything he wanted to read, and he was immersed in reading material. It was clear that he loved books, but there were other odds and ends scattered around the store.

Here I am, age 12, with a picture of a microscope just like mine.

I came across an old monocular microscope covered in dust, abandoned in a corner and buried by books. The shopowner saw little value in this device. Its faded price tag said it was available for a few pennies. Being a researcher at heart, I had determined using a piece of paper and light from the window, that this wonderful microscope had remarkably good optics.

I now started to explore the microscopic fauna and flora of my favorite ponds. Amebae pursuing their prey, hydra looping along a surface, the beauty of *Spyrogyra*, the inner life of round worms and other beautiful *animalcules*. Many were recorded on black and white photographic film to be printed in my darkroom, presaging a long career as a microscopist. Anton van Leeuwenhoek would have been proud of me.

Finding an old book of stains for microscopic sections, I tested my skills as a histotechnologist. This is the study of techniques used to reveal tissue structure in the light microscope. Special stains show different tissue components, such as starches and lignin. Not possessing a microtome, I cut slivers of plant stems with discarded razor blades, and stained them with dyes purchased at the chemical supplier. More curious looks from sales staff for this odd 14-year-old kid.

These boyhood investigations bore fruit many years later, when I took charge of a large histotechnology laboratory.

I Become An Electron Microscopist

While running the neuropathology diagnostic service, I would also receive sick sheep, sometimes *in extremis* (almost dead), and some were dying of a disease known in Britain as cerebrocortical necrosis (CCN). Here I saw a research opportunity.

I had heard of transmission electron microscopy (TEM), which would allow me to take a closer look. I also knew there was an electron microscopy department that largely served the virology department. Preparing tissues for TEM is tricky, involving expensive equipment. I approached the TEM department staff, with a request that I study CCN samples in their electron microscope.

I cannot say I was welcomed with open arms. They were a close knit group of three, two men and a woman, who looked at me as though I was just another deranged young scientist. They explained that my brain samples would have to be fixed in a particular way, perfused with glutaraldehyde and embedded in plastic. They would then cut extremely thin sections on their automated ultratomes, place the sections on little metal grids, and prepare them for me to examine in the electron microscope.

I persevered, collecting samples from brains of normal sheep and those dying of CCN that I had perfused correctly with TEM fixative. I took these samples to the TEM department, where they were greeted with a sigh. Several weeks later, they informed me that they had a sample ready for me to examine.

I asked how the sample was oriented, and they looked at me with confusion. They were used to looking for viruses in tissues, for which tissue orientation was irrelevant. I was a whole new deal. One year of struggles later, I developed a way to embed tissues for TEM that permitted recognition of orientation in relation to the levels of the cerebral cortex (outside to inside).

I needed hundreds of sections to do the job.

They looked askance at my embedded blocks, informing me that they were no good. They also said that they could only cut two samples per day, so I could forget my hundreds of samples. On receiving a single grid from my samples, and looking at them in the electron microscope, I realized that I was onto something useful. But how to get enough grids into the scope?

I was used to light microscopes, with a 1,000x limit of magnification, but here was a microscope that could magnify sections of tissues up to 100,000 times. Just what I needed to understand the brain disease I was studying. The problem was the people who cut sections for this fancy microscope.

Every time I suggested an improvement in the way my tissues were handled, they would say, *We don't do it like that, because it doesn't work.*

But I was raised by my mother to question authority. I didn't blindly accept these statements, which turned out to be excuses. I decided that *I would show them.* I would find a way to cut one hundred samples per day, with less sophisticated equipment.

I located an inexpensive, manual ultramicrotome in a scientific catalogue. I then explained my dilemma to my boss and asked if we could buy it. He was only too aware of the poor sample turnaround from the TEM department. With a glint in his eye, he said I could give it a go.

It took me months to master this unsophisticated machine. I had to ask the TEM staff if I could use their glass-knife breaker. These plastic sections are cut on glass or diamond knives. They wanted to know why, and came to my office to see my cheap ultramicrotome. Their machines were top of the line, electronically controlled, and sitting on vibration isolation tables, with top-of-the-line diamond knives. They took one look at my contraption, laughed and said sure I could use their glass knife breaker. They could not wait to watch my failure.

Even though these people were nice, they had been in control of the TEM department for too long. They were stuck in the past, when TEM was only used for finding evidence of viruses in tissues.

For months, I worked away embedding tissues, breaking glass knives, cutting sections onto a tiny water bath attached to these knives with paper tape. I learned to pick up the delicate, ultra-thin sections, onto tiny metal grids. The grids were placed in a grid box with room for 100 samples prior to staining for the scope. I finally mastered the cutting of long ribbons of perfect silvery gold sections, transferring them safely to my grid box. With their fancy equipment, they said they could only cut two samples per day. My goal was to prove, beyond a shadow of a doubt, that this daily two-sample limit was a lie.

Time for the showdown. I worked for about seven hours, cutting 98 individual samples, almost filling my grid box. Then, knowing the TEM staff were in their den, chatting about politics as usual, I cut a perfect ribbon onto my water boat. I then went quickly over to the TEM suite, and explained that I was having trouble cutting sections of adequate quality, *the set up*. Could they take a look and give me some guidance, please? Over they came to my office to have a laugh.

They each took one look at my section ribbon, and remained silent. This was followed by the knockout punch: I said I had hoped to cut 100 samples that day, but only had 98, and I needed to go home. I came by the TEM facility the next day, and not a word was said.

This silenced them, but it made them a little angry. *The jig was up*. They could no longer pull the wool over innocent eyes. With tact and persistence, they became more accepting of my new ways and grew to respect me. From then on, they provided me with as many samples as I needed. They accepted my new embedding procedures, designed for pathology rather than virology, which opened that TEM department to other pathologists.

We eventually became good friends, as I had earned their respect. On leaving the Moredun Institute several years later in 1975, they presented me with a gift of the 2-foot square photo below. It sits on the living room wall in front of me as I write this in 2018.

The gift I received from the Moredun Institute TEM department staff, in 1975. A perfect electron micrograph of a spore of the bacteria I isolated from sheep, B. thiaminolyticus.

They had worked hard and long to make that gift. You can change hearts and minds, but it takes a little effort. It also helps have a little obsessive compulsive disorder.

Science Survival Tip No. 27: *If people get in your way, beat them at their own game, to earn their respect or at least to make them fear you.*

This goes for anything. Such as, hybridizing a gene expression array, running a Northern blot for messenger RNA, building a mathematical model, exploring computational fluid dynamics, doing a necropsy on a rat, making up buffers, running statistical analyses, working with HPLC systems, cannulating thoracic ducts in mice. All of these I did. Not well in every case, but I knew exactly what my staff were doing.

My research on CCN became the subject of a doctoral thesis that I submitted to the veterinary pathology department of the Royal (Dick) School of Veterinary Medicine, Edinburgh University, Edinburgh, Scotland.

1975: I Survive my Ph.D. Defense

My doctoral thesis sported the title *Studies of Ovine Polioencephalomalacia*. The name I chose for this disease was not appreciated by the examining committee, but it was the name used in the United States. Furthermore, post-war anti-American feelings were still running high in the British landscape.

Oversexed, overpaid and over here!

On examination day, as a boyish young man, in my late-twenties, all ginger hair and enthusiasm, I arrived on time at the veterinary school. It is an intimidating building, which has now been converted into an art center and a nice bar that is decorated with veterinary paraphernalia.

Back in 1975, I knocked on the door of the office assigned for my examination. On receiving no audible response, I entered a somber, wood lined, dimly lit office. Awaiting me were three inscrutable men. They wore serious expressions and the standard garb for veterinary research examiners: Tweed jackets, university ties, and somewhat aggressive postures. They instructed me to take a seat. I wondered how this was going to go, given the evident sense of mild hostility in the air.

Each member of his examining committee was respected in their field of research. Pathology, biochemistry, and bacteriology. These were disciplines I had employed extensively in my doctoral research. These three men held in their hands the power to grant or withhold Ph.D. after my name. A doctorate is a critical meal ticket in the world of research.

My mother was tough, and she told me many times that *you can fight city hall.* Meaning, stand up to power and you will earn their respect. Despite that, I was still nervous.

In the university standard gold type, on the front of my two-volume thesis, was the source of the tension. A copy was in front of each examiner, with many paper markers inserted into each copy. At the time, the British name for the disease that was the subject of this thesis was *Cerebro-Cortical Necrosis* (CCN). My defense opened with an immediate attack on the selection of the name, *Polioencephalomalacia (PE)*. A better name for this disease, in my not-so-humble opinion.

The subsequent intellectual sparring wasn't based on feelings. It wasn't based on pro-North-American sentiment. I had no emotional attachment to the name I'd chosen. I was attached to reason. Always have been. I had opted for the name *PE* because it was a better name. The brain damage associated with this disease is not confined to the cerebral cortex. The decision was based on logic, and knowing more about the subject than anyone else in the room.

When you sit for your Ph.D. exam, you had better know much more about the subject of your thesis than your examination committee, and anyone else in the world for that matter, or you will be in trouble.

I prevailed, and on that day, in 1975, I became Dr. Kevin Morgan, B.V.Sc., Ph.D.

The author with his doctoral thesis, over 40 years after he earned it.

I was finally a real researcher, even though I had been one all my life. This experience imbued me with a sense of the importance of *lexical semantics,* especially when it comes to the names of diseases. Interestingly, this turned out to be critical many years later.

Retired from my research career, during a lecture in 2010, I endeavored to tackle a tricky problem. So-called plantar fasciitis, which led to the publication of this book, a few months ago.

Science Survival Tip No. 28: *It ain't what you don't know that gets you into trouble. It's what you know for sure that just ain't so.*

— Mark Twain

How do I know it's significant work? It was called:

- Junk science, by a scientist.
- Based on BS, by a podiatrist.
- The worst kind of garbage, by another podiatrist.
- Its working hypothesis was banned from a site run by two surgeons. Why would you ban an hypothesis? Simple, if it threatens your business income and you have porous ethics.

Clearly I have not quite retired from science. I am still in the game at 75, because I am in my element.

HOW SCIENCE IS ACTUALLY DONE

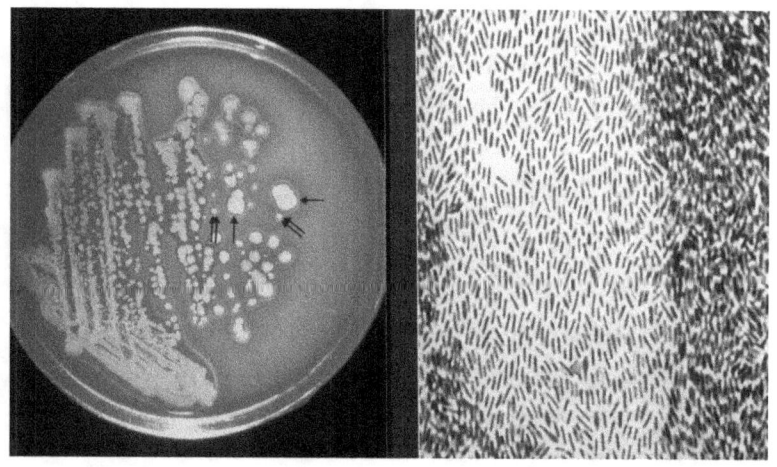

> *Chance favors the prepared mind.*
> — Louis Pasteur

Science Survival Tip No. 29: *Scientific disciplines are only human tools. They come and go - when did you last consult your alchemist or necromancer, for instance?*

I am pretty sure the cells in your body don't say to themselves,

Time for some fluid mechanics, and how about a bit of chemiosmosis and quantum physics, while I'm at it? The cells of your body do everything at once. Chemistry, physics, molecular biology, structural engineering, mathematics and so on, all at once. But it takes chemists, physicists, molecular biologists, engineers, and mathematicians, respectively, to master each of these skills in our "big" world.

This fragmentation of reality can create communication problems. I once attended an advanced mathematics lecture with an audience comprised almost entirely of mathematicians. I was convinced that the speaker was the only person in the room who *really* understood what was going on. The question-and-answer session that followed confirmed my opinion.

With such fragmentation of our comprehension within a single field of study, how are we to make progress when everything covered by every scientific discipline is seamlessly connected in the organisms we study? I suspect that we do need specialists, but we also need people to exploit their knowledge in a way that connects the disciplines.

We clearly need some Jacks of all trades, masters of ONE. You have to master one trade, or you won't get a seat at the table.

Unbeknownst to me, I became a *Jack of multiple trades, master of one*. I mastered pathology, even though one never stops learning one's trade. I certainly have the kind of proof of my mastery exploited by *The Wizard Of Oz*:

- A doctorate in neuropathology.
- Fellow of the Royal College of Pathologists. Well, Ex-Fellow, as I stopped paying my dues years ago, but I was a fellow until I left in England.
- Diplomate of the American College of Veterinary Pathologists.
- Lots of articles with pathology content.

Pathology lends itself to the incorporation of multiple disciplines because you are led by the nose by the nature of the diseases you investigate. This requires a lot of studying, leading to the ability to talk to experts in other fields in their particular language (math, biochemistry, molecular biology, physiology, etc.). This ability has to be sufficient for them to (a) take you seriously, and (b) become collaborators in your research efforts.

Essentially, you need a trade and willingness to learn whatever you need to learn. Or, more precisely, what the problems you are trying to solve demand that you learn.

1973-74: Isolating Ovine *Bacillus thiaminolyticus*

The subject of my doctoral research in sheep diseases that I studied in Scotland in my 20s, was polioencephalomalacia. It was suspected that the condition was because of the presence of an enzyme in the digestive tract of affected sheep and cattle that destroyed vitamin B1, or thiamine. This enzyme is called thiaminase type 1. It not only breaks down thiamine, in the process it also converts the thiamine into a molecule that competes with vitamin B1 biochemically, thus further impairing the vitamin's functions in energy metabolism, amongst other vital things.

One question I decided to tackle was to track down the source of this enzyme in the rumen (stomach) or feces of sheep.

Problem: Where does the thiaminase in the rumen and feces of affected sheep come from? Bacteria, protozoa, or something else?

First choice. Bacteria, as there were already two known sources of enteric thiaminase type 1 in other species, including humans, and both were bacteria. This is when I started going around in circles.

What I needed:

- An assay for thiaminase type 1 so I could track it to its source, and a biochemist to help me set it up.
- A bacteriology laboratory with people willing to train

me in bacteriology and help me solve bacteriology problems.
- A source of rumen contents and feces of affected sheep.

Solution — networking.

1. I started asking around for people with these skills. A friendly biochemist, Neville Suttle, set up a radioactive assay for thiaminase type 1. He proceeded to train me in the assay procedure, how to safely handle radioactive carbon-14, and in the use of the scintillation counter. One down, two to go.
2. As I needed a second advisor for my doctoral work, and I needed bacteriology training and resources, my primary advisor sent me to the veterinary school field station to meet Gordon Lawson. Gordon, and his scary (strict and really great) technician, Penny Wooding, took me in hand. Their training converted me into a not-too-useless bacteriologist. Two down, one to go.
3. I then contacted local veterinary field services where I met Karl Linklater. He not only provided an endless supply of feces from CCN outbreaks, we even published one field study together [13]. Rumen contents were obtained from flocks of sheep showing symptoms or dying with the disease. The sick sheep came into our facility for me to necropsy, as a staff pathologist.

Done!

Now the logistics, which are rarely, if ever, mentioned in scientific articles.

Twice a week, I would take samples of feces or rumen contents from our freezer in Edinburgh, drive 10 miles to the bacteriology lab at the veterinary school field station (only 2 miles from my house, fortunately), and place them in liquid broth for aerobic and anerobic culture. The next day, I would take samples of the broth from these cultures, filter sterilize the samples and take them back to our institute for the thiaminase type 1 assay. If there was a strong thiaminase signal in any of the broths, I would return to the bacteriology lab in the evening, and plate out such broths onto a range of agar plates, and then plonk them in the incubator or into anaerobic jars.

For 3 days, I would examine these plates for interesting bacteria. I would collect individual colonies and grow them up in broth, take the broth back the 10 miles to our institute, and then run the thiaminase type 1 assay. I was trying to find bacteria that made the thiaminase type 1 in my samples. Of course, I am still running the neuropathology diagnostic service, attending biochemistry lectures, and running other, related and unrelated [1,4,5,6,9] experiments.

If you want to flourish as a young research scientist, you have to hustle.

I continued the cycle of (1) analyzing feces or rumen contents, (2) culturing positive samples, (3) isolating bacterial colonies, (4) culturing them in broth, (5) analyzing the broth, (6) re-plating the broth to ensure pure bacterial samples, (7) growing them up in broth, (8) analyzing this broth for thiaminase type 1 activity.

And for about 18 months, any time I purified the bacterial colonies the thiaminase type 1 signal would vanish.

What the hell was going on?

Maybe the enzyme was made by something other than bacteria, and was simply contaminating my samples and broths? I was considering moving on,

When,

just like Alexander Fleming and his discovery of the *Penicillium notatum* mold, the source of penicillin, I left some plates in the incubator longer than 3 days, over a holiday weekend. I completely forgot about them for about a week. On returning to the bacteriology lab, I went to toss these dried-out, cracking agar plates, but I took a quick look at them, anyway.

There, on the surface of one particular blood agar plate, was something new. I had been staring at these growths for over a year, remember. I had seen every kind of bacteria you can imagine. I took my platinum loop, poked the new growth, and found that it was sticky and adhering to the loop. *Something new.* I spread a small sample on a glass slide, stained it with Gram's, to find millions of Gram positive, bacterial rods, many containing large endospores.

I had seen this somewhere before. *Damn.* I then assumed that I had grown up the *Bacillus thiaminolyticus* I had purchased from the Type Culture Collection, ages before. It had looked just like that, spores and all. I must have contaminated the incubator or lab with some of these spores.

Solution: Destroy all of the samples, clean the lab and incubator, and start again with one little change. Examine the plates about a week after plating samples and see if a different strain of this particular organism was in the samples. Then, check to be sure it was not the Type Culture Collection strain.

I did that, and within a week I isolated a similar, but not identical organism. *It was not a contaminant!* It was the first (to be reported) ovine strain of *B. thiaminolyticus* [7].

I re-cultured and purified this organism, grew it up in broth, and analysis revealed a strong thiaminase type 1 signal. On further purification of this bacterial strain, the thiaminase signal remained consistent. I showed my results to Gordon Lawson, who then worked with me to optimize the isolation process. This resulted in the rapid identification of 10 new strains, each having its own personality. Stickier, or less sticky. More, or less active spore production.

Gordon then instructed me in the process of classifying these organisms, which were accepted as a new ovine (sheep) strain of *Bacillus thiaminolyticus* in the Type Culture Collection. These bacilli are facultative anaerobes, meaning they can grow with or without an atmosphere containing oxygen. The story became more complex, as another lab isolated a strictly anaerobic, thiaminase type 1-producing organism, *Clostridium sporogenes*, from similarly affected sheep at about the same time.

Almost none of my "running round in circles" (collecting and analyzing gut contents, work in the radiation and bacteriology labs all hours, and round and round again) can be gleaned from the related publication [7], or the excitement in our bacteriology lab, as Gordon and Penny had just isolated new strains of *Campylobacter sputorum subspecies mucosalis* from pigs.

Those were exciting times that were invisible to the outside world!

At the same time, I was carrying out studies of a model of PE, amprolium toxicity [3,8]. This work, combined with the neuropathology diagnostic service, exposed my young mind to:

- Pathology
- Electron microscopy
- Analytical chemistry
- Biochemistry
- Hematology
- Histochemistry
- Nutrition
- Ruminant science
- Field studies
- Statistics
- Toxicology
- Virology
- Sociology

Sociology?

Yes! I couldn't have done any of this, without the help of Dick, Karl, Neville, Gordon, Penny, and multiple technical and animal care staff at the Moredun Sheep Diseases Research Institute.

That is how science is *actually* done.

READING AND PUBLISHING

Science Survival Tip No. 30: *Read well, to write well.*

I was a late bloomer. I did not learn how to read until the age of 8 or 9 with the help of my older sister, Janet, starting with *Ananci was a Spider*! I had no idea about the roots of this old tale, I just found out that she was bold and black. Once I caught on, however,

I became a voracious reader. This is what saved my ass, educationally.

I mentioned failing the 11 plus, at 11 years old. Usually you were done for unless you had rich parents or were in the "right family." Yes! England was a class system, like it or not. It probably still is to some extent. The world seems to change rapidly, but many things stay the same.

On learning that I had failed the 11 plus, I remember it as one of the only times Mum hugged me. I was distraught, and Mum had realized it. She assured me that she would find a way, money or no money, and she did. Somehow, Mum paid for me to go to a small private school run out of someone's home about a mile away. Here, the class size was no more than 10 kids, not the 50 I was used to. I *actually* got to speak to the teacher. They helped me with math and reading, and I took the 11 plus again, by which time I was 12 years old.

I failed again, but apparently only just. The school assured my Mum that I might just scrape into the grammar school through an interview. Each year, some of the kids who passed the exam, and thus had a place in the grammar school, had to move away or even died. Dying was a little more common in post-war England. Anyway, I was offered to chance to have an interview and compete for one of the few vacated grammar school seats. Remember, no grammar school, no university. It was the way things were.

I was a shy kid who read a lot. I would scour the secondhand book shops, where you could by a book for a penny. I was pretty good at spotting dropped coins in the street. Once, I found a half-crown. A small fortune, which could buy a pile of books (and bread rolls or broken biscuits). By the time I was 12, I had read all of the books by Edgar Rice Burroughs that I could lay my hands on. All of Tarzan, and the Mars series. I was also halfway through the works of Dickens. My favorite is still *A Christmas Carol*.

My private school teacher ushered me into her office with words of encouragement, where I met the interviewers. These memories are vague. I remember a man and woman looking at me severely and peppering me with questions. They asked me if I liked to read. I guess they had expected me to say I preferred soccer, but I regaled them with Tarzan and Dickens. They seemed to pick up on Dickens, asking about the details of those remarkable tales, which led to important social changes in England at the time. Dickens was an agitator for the rights of the poor.

I got the position and went to St. George's Grammar School. The journey from there on was like that of Harry Potter, without the magic. Or without the wands, anyway.

St. George's Grammar School, Bristol, England.

I arrived at school, nervous on my bike, as my older brother was waiting for me. He was a school prefect, and as I came through school gate pushing my bike, he gave me a detention. I asked why, and he explained that it would do me good. That was messed up, but for some reason I just accepted it as the way things were. However, I was not completely well-adjusted, either.

My class was in a trailer with about 30 other kids who had just passed the 11 plus. Our desks had inkwells, even though we now had fountain pens. For some reason, knowing a bit of chemistry, I decided after a few weeks to fill an inkwell with powdered sulphur and light it. The classroom filled with highly irritant sulfur-dioxide, and was evacuated. I was threatened with expulsion, but for some reason was given a second chance.

I guess my start in grammar school was promising for a scientist. I had an interest in experimentation.

That said, in my teens, I was happiest in my small dark room at home. I afforded the expensive hobby of photography by earning money at a local hotel as a handyman and working as a laborer on a pig farm, which was a 12-miles bike ride from our home. This was hard-earned money, so I soon discovered ways to work around certain expenses.

For instance, out-of-date-film was always as good as new film, and the local camera storekeeper would often save it for me for free. An interest in chemistry at school led me to make my own photographic developer and fixative from raw materials purchased in bulk at a chemical supply house. The staff would look at this odd kid in his school uniform with bemusement as I purchased a 20-pound bag of ammonium thiosulfate, for instance.

My early experience with the chemistry of photography played an important role in those studies of fluid mechanics years later.

In 1958, I worked on that pig farm for an entire summer vacation to earn the money to buy a decent camera. If you have never painted the inside of a wooden pig house with creosote on a hot summer's day, you have never lived!

I delved into books on optics and the chemistry of photography. I now realize that I was more interested in these things than the photographs themselves. I loved to watch the images from my camera come to life on paper in the developer tray. Such are joys long gone with the advent of digital photography, but they are coming back. Would you believe it?

———

I recently attended the graduation of my eldest grandson, Tyler, in

Kyoto, Japan. During this trip, I visited my step-daughter, Jess, in Sidney, Australia. There, I noticed two cameras in a shop window display. They were nearly identical to the ones I owned in my teens in the 1950s. They were perfectly preserved, but a far cry from modern, digital cameras, or even our phone cameras.

Two old cameras I spotted in a shop window in Sydney, Australia, in 2018.

I enquired with the owner of the camera shop as to why he kept these old relics, and he informed me that the old ways were coming back. Young men, in their early 30s especially, were buying roll film. Some were even setting up darkrooms of their own. I asked one such customer why he was going to all this trouble, when digital photos are so convenient. He replied, "I love the excitement of waiting to see what the photos are like."

This exchange reminded me of my doctoral research, and the use of paper chromatography. You get to see the separation of the chemicals on the paper. It is not the same when a high pressure liquid chromatography system, linked to a mass spectrometer, spits out a mass of numbers.

I mentioned my 1970s paper chromatography experience to one of our post-doctoral students, Matt, in the late 1990s. We were planning a biochemistry experiment, and his jaw literally dropped.

Matt looked at me and said, *You mean all those glass jars? I saw some in a science museum, recently.* It made me feel like I might belong in a museum, myself.

Survival Tip No. 31: *Consider playing with the old methods, or at least read of their history.* For instance, puzzled by statistics? Read *The Lady Tasting Tea,* by David Salsburg.

As a teen, I took photos of animals on the farm where I worked. One particular image of pigs feeding at a trough was accepted for a photographic show in London. This seemed like an exotic location far from Bristol, and I could not afford to go there. I later received a letter saying that my photo was well received.

This gave me a much-needed boost. *Maybe I am good at this after all*.

If you find you're an outlier, appreciate it as an advantage for your science career.

Science Survival Tip No. 32: *Don't worry about people stealing an idea. If it's original, you will have to ram it down their throats.*

— Howard H. Aiken

Using Gravity to Your Advantage

I rapidly realized that to survive in scientific research, Ph.D. after your name can be critical. However, as this book demonstrates and as Feibelman explains in his excellent book, *A PhD Is Not Enough*.

Sheep affected with PE fall around, tumble over, and die in seizures, while suffering a great deal in the process. These symptoms are because of gradual softening of the gray matter in their brains. Yes, sheep have gray matter. *Lots of it!* We have a tendency to underestimate the intelligence of non-human animals. Especially sheep, I'm afraid.

Sheep are wonderful creatures, and great mothers. Based on the structure of their brains, I suspect that they enjoy fascinating dreams. How else could they stand in the heather of the Scottish hillsides, year after year, happily chewing the cud and raising lambs? Furthermore, they can provide us with both food and clothing, permitting humans to survive in otherwise uninhabitable regions of the planet.

The underlying cause of PE is a problem with vitamin B1, thiamine. This vitamin is important for our (human and sheep) bodies to turn food into energy. The brain needs lots of energy, consuming about 25 percent of all the energy that our bodies produce. Beri-beri is the nervous dysfunction associated with vitamin B1 deficiency in humans. We are more like sheep than most people imagine.

My most recent article was bulky because I had employed a range of diverse disciplines, from the study of blood (hematology) to brain biochemistry.

I could not submit the paper to a journal for publication without the signature of the department head, that being Dr. Dick Barlow. The 50 pages of text were typed on my Corona portable typewriter, with printed photos glued onto the pages. No computers in those days. Not for word processing, anyway.

Photographs were created using the wet chemistry methods of the day.

Digital photography was only a glint in the eye of Steven Sasson at Eastman Kodak, back then. Black-and-white photography was one of my passions as a teenager, financed with money earned on a local farm and in a hotel.

Mastering the pathology department darkroom equipment was no problem. Everything you learn comes in handy one day.

Dr. Barlow accepted my precious bundle, saying, "Good work, Kevin! I'll get it back to you in about a week," which he did! Covered in red scrawl. "Not bad! Not bad at all. For your second publication. But it's too long. Much too long! I've made some suggestions."

The process was repeated about three weeks later. To my dismay, more red scrawl: "Still too long. Kevin. Much too long."

On the third attempt: No red scrawl, but, "Still too long! You're going to have to shorten it, or break it into two papers! I don't know how you're going to do it, but it's still much too long to publish. Sorry!"

You have to appreciate my dilemma to understand the challenge that I faced. Publications are the currency of science. Talk is cheap, but you have to walk the walk with publications. Difficult to create. Based on valid data from well-designed and relevant experiments, and containing a detailed interpretation of the findings. Demonstrating a clear contribution to human knowledge. Then there's the peer review and journal editorial process. If you are challenging an established authority, this process can take years.

Believe me, I have been there!

My wonderful creation on mushy brains in sheep, now much-improved thanks to Dr. Barlow's input, would not work as two papers. *I knew it.* I had removed everything I could. I liked it as it was. It was one, long, interesting and informative story.

What was I to do?

Take your mind back to the word-processing technologies of the 1970s.

Survival mechanisms kick in. I do not like Dr. Barlow's two papers advice.

Science Survival Tip No. 33: *There's always a way to overcome roadblocks, you just have to find them.*

A lightbulb went on. I have no idea where the idea came from, but I can guess. A book or a movie. I re-typed the manuscript on airmail paper, as this paper is designed to reduce shipping costs. It weighs about one-third that of ordinary white vellum. I also reprinted the figures on the lightest of photographic paper.

I look back on those days from the vantage point of the more-advanced technologies of today. We had a closer relationship with each photograph. Is this worse? No, it is different. At least I did not have to make etchings for my publications. The Victorians would have said I had it easy!

I think that we may have had a somewhat better relationship with our bodies, in the absence of constant sensory input, from ear buds and the like.

Back to 1973

I weighed the lighter, floppy paper in my hand, thinking,
Will this work?
Will Dr. Barlow notice?
Is it honest?
Is it science?
Dr. Barlow accepted the document and did not notice anything was amiss. Two days later, he called me into his office with a smile and handed it back, unmarked. "Great work, Kevin. I enjoyed reading the shortened version. That's much better. You can submit it for publication!"

The journal accepted and published my lengthy article with

little to no changes. A rare event, indeed! Of course, it was submitted to the journal on regular paper.

Dr. Barlow's brain had employed the weight of the bundle of paper as a measure of the manuscript's length. This was subconscious, as he was unaware of my subterfuge. Once it had "gone to press," I told Dick (no longer Dr. Barlow) what I had done, over a beer.

The Robin's Nest Pub, on Gilmerton Road, in Edinburgh, Scotland, where I really learned how research is done back in the early 1970s.

No, I waited until the second beer. We were at our local pub, The Robin's Nest, on a Friday night after work to celebrate the paper's acceptance by a good journal.

Dick revealed a flash of anger as he learned how I had tricked him by fooling his senses. Then, he laughed out loud and bought me another pint to toast a successful subterfuge. After all, I had provided another publication for the department!

What had I learned from this adventure?

Even logical scientists are at the mercy of their subconscious, making them open to manipulation.

If you're still in doubt, read *The Art Of War* by Sun Tzu, be honest about your actions, and listen to one of the greatest martial artists of all time (in my opinion):

Science Survival Tip No. 34: *Adapt what is useful, reject what is useless, and add what is specifically your own.*

— Bruce Lee

Yes, I know we do not use airmail paper today, with cell phones, FaceTime, Twitter, Instagram, email and all the rest, but the same principal applies as you learn how to become a successful scientist today. Technology has changed dramatically since 1973. *Human nature evolves more slowly.*

My Favorite Data

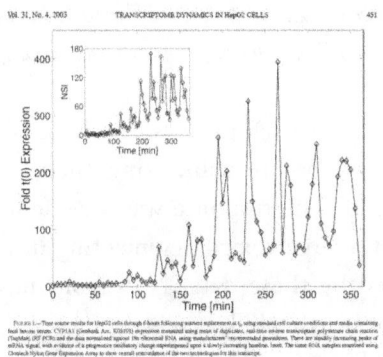

I loved these data, but no-one else seemed interested [124].

While running a large molecular biology lab, I took advantage of our facility budget to study the impact of nutrition on cell cultures. I have always been interested in food, and how I ran the lab and the research we chose was at my discretion. I decided to do a dense time-course experiment: samples every five minutes for six hours of cell cultures after changing the media (cell food and waste disposal). The cells were analyzed with ClonTech Arrays®. The results were remarkable, with transcripts bounding around, or ticking like a clock.

These data were mind-boggling, leading to my interest in the mathematical modeling of transcript dynamics. These results were one of my best holiday gifts as a scientist. I loved these data — beware. I found it possible to regenerate this graph with three trigonometry functions, meaning it may be the result of the activity of only three transcriptional regulators.

This work on transcriptome dynamics was rejected by two major journals, as being, (a) unconfirmed because we did not use Northern Analysis - we confirmed the array results with a superior, but at that time yet to be widely accepted, technique called Taqman™, and (b) as being of little interest.

Whatever!

Science Survival Tip No. 35: *Just because you love your data, do not expect the world to be equally impressed. However, one discovery can change the world, even if you are not there to see it change.*

This was a fascinating process. After modeling these data mathematically, and before submitting them for publication, I took it on the road. The response was a deafening silence and an odd statement at a large toxicology meeting in England. Remember, I was excited by these data. Not about how they made me look, which is not worth a hill of beans.

I was excited because I was now seeing cells more as complex sets of electronic equipment rather than Lego blocks. This was all thanks to the re-invention of cloning by Kary Mullis. The stories of his conflicts with corporate suits are fascinating.

Here I was, in my 50s, presenting the dynamic responses of a few hundred transcripts to media replacement using gene expression array technology. This technology was leaving the gold standard of Northerns in the dust. I did one once, just for fun. It was a pain in the ass, yielding mediocre results for a single transcript. It did increase my respect for the molecular biology researchers who had gone before.

During the talk, I proceeded to explain the dynamics of selected transcripts in our data in relation to biological clocks, glycolysis, oxidative phosphorylation and the urea cycle. *I loved this stuff*, and by then I was an experienced speaker.

At the end of my talk, there was another deafening silence. Then an old gentleman (probably younger than I am, today) slowly stood up in the middle of the audience. I suspected that he was the feared *Grand Old Man* of that society. He stared at me for a moment, raising memories of Dr. M, and said,

"Dr. Morgan, your data frighten me to death."

Then he abruptly sat down.

The chairman quickly moved on, to introduce the next speaker. Was I disappointed? No, I was delighted. Such a response is the sure sign of a paradigm shift in the wind. It may not be today, tomorrow, or a hundred years from now, but it is a sign of change to come.

Consider the equation for an ellipse.

A mathematician friend told me of this story while we were discussing the value of citation index and the challenges of ignored ideas. Here is an extract from an article on this important subject that finally led to the description of the movement of planets in our solar system by Kepler and Tyco Brahe.

From *Conic Sections in Ancient Greece by Ken Schmarge, History of Mathematics Term Paper, Spring 1999*

The knowledge of conic sections can be traced back to Ancient Greece. Menaechmus is credited with the discovery of conic sections around the years 360-350 B.C. These curves were investigated by Aristaeus and Euclid. The next major contribution to the growth of conic section theory was made by the great Archimedes. Apollonius, on the other hand, is known as the "Great Geometer" on the basis of his text Conic Sections, an eight-"book" (or in modern terms, "chapter") series on the subject. The first four books have come down to us in the original Ancient Greek, but books V-VII are known only from an Arabic translation, while the eighth book has been lost entirely. Pappus, who lived about 300 A.D., furthered the study of conic sections somewhat in minor ways.

After Pappus, however, conic sections were nearly forgotten for 12 centuries. It was not until the sixteenth century, in part as a consequence of the invention of printing and the resulting dissemination of Apollonius' work, that any significant progress in the theory or applications of conic sections occurred; but when it did occur, in the work of Kepler, it was as part of one of the major advances in the history of science.

I do not think Apollonius's work is mentioned in citation index. I bet he did it because he found it interesting. That is the best way to do science, and you will know when you are doing that.

You will be *in the zone!*

STORY WARS

Science Survival Tip No. 36: *Those who tell (and live) the best stories will rule the future.*

—Jonah Sachs

Your public speaking and storytelling skills can make or break your career. The basic rules are simple:

- Make them laugh.
- Make them cry.
- Never make them bored.

I lived in terror for several weeks before my first scientific talk. I then went to Toastmasters for about four years, and that made my public presentations more bearable, but never fun. Then, one evening I was listening to the violinist, Yehudi Menuhin, talking on the radio. The interviewer had asked him if he was ever nervous. He replied with this short story:

"I was nervous early in my career, and I wondered if I would ever be comfortable up there. After I'd been performing for about twenty years, I was invited to play the solo part of a particularly difficult violin concerto. The concert hall was packed with over two thousand people. I was provided with a chair in front of the stage.

This particular piece has a long continuo, before the soloist enters. I sat quietly on my chair as the music started. It was truly delightful. I enjoyed listening, but a niggling question appeared in my mind. Something is missing. I wondered what it was. I knew this music so well.

Suddenly, I realized the missing part was me. My solo part. I was so enjoying the music I forgot that I was supposed to be performing. The first violin was gamely attempting to fill in for me.

It was then that I learned that I was no longer nervous. I'd realized that the music isn't about me, it's about the music."

I found this story hard to believe, given my terrible stage fright, but the years went by, and about 20 years later I was attending a scientific meeting in Dallas. I was at the back of the main hall, packed with about 600 scientists. I cannot remember the focus of the meeting, as I only remember being at the back of the room quietly talking to a friend between the lectures. One had just ended, and another was about to begin.

My friend and I were engrossed in conversation when my friend said, with some surprise, "Kevin, they keep calling your name. What's that about, do you think?"

I smiled, remembering Yehudi Menuhin, and said, *I'd forgotten, I'm the next speaker,* and calmly walked up to the podium.

The story isn't about you, it's about the story.

A wonderful life guide for living well.

The Turk

I did enjoy *Le Petit Prince* by Antoine de Saint-Exupéry, and I was struck by one particular story. People would sometimes comment on my mode of dress, especially in corporate hallows. I would wear whatever was comfortable, generally shorts, T-shirt and running shoes. In fact, one human resources staff member reminded me in passing that I was not following the company dress code, and I had been seen on the security cameras, skipping down the main administrative corridor. This was stated with a mixture of admonition and awe.

Our cat, Cat. Approach with caution.

To you, from Cat — **Science Survival Tip No. 37:** *It is alright to wear the same clothes everyday, remember to bite the hand that feeds you, and never put on the golden handcuffs.*

I replied, to the human resources lady, not Cat, that I had been hired for my brain, not my dress sense. I cannot think in a suit, or pipette radionuclides like that, either. *I have to be comfortable to function.* Then, the picture of the Turkish astronomer in Le Petit Prince, and his drawings of the *astéroïde six cent douze*, popped into my head.

The stranded pilot tells the story of the Turk's great discovery of this asteroid. He presents his discovery at an international astronomy symposium, but because he is wearing Turkish dress, including a Fez, no one takes his work seriously.

He returns home to continue his studies, and then there's a coup in the country. The new president demands that everyone abandon their traditional clothes for Western dress. The astronomer, now in a smart suit with no fez, presents his discovery of the asteroid B612, yet again. The only difference is his clothing. Now his discovery is met with considerable interest and applause, just because he is wearing a suit.

I have an unfair tendency to distrust people in suits, in scientific meetings, at least. I always wonder if they are hiding something. That said, I wore a suit during the first twenty years of my career, when speaking at scientific meetings, until one day in Texas.

Dallas, 1995

I was invited to give a lecture at a conference on cell replication in toxicology. My topic was cell proliferation in the nose of the rat and it's relationship to cancer. I was still a neophyte when it came to public speaking, despite four years of Toastmasters training and a growing number of lectures under my belt.

My problem: I had yet to completely get over myself. It is critical to realize that it is not about you, it's about the story you are telling. Get that right, and all will be well.

My lecture was scheduled for 9:30 a.m., just before the break. The room was full. About 200 people, and many had paid to attend. These lectures could be logged as continuing eduction. I was listening to the talk before mine a little nervous as usual. Next to me was a friend and fellow scientist. A peer, but also a pathologist. We were at the back of the room, as I like to be near an escape route. I can only listen to so many lectures in a day.

As the previous speaker was wrapping up, I said to my friend, "Hey! Jeff, do you think that they'll be offended that I'm not wearing a suit?" He replied, *"I'm sure they won't care a bit."* That morning, with the temperature sitting at around 105 degrees, I could not face the thought of a hot suit. I had noticed that almost all the other speakers were wearing suits.

Up I went, suit-less. Once I was into the lecture, enjoying my subject matter, I forgot all about it. I did not detect any kind of reaction from the crowd.

Once it was over, I returned to my seat next to Jeff, who immediately leaned over and whispered in my ear, "Kevin. See those two guys over there? Well, when you walked up to the podium, the guy on the right said to his companion, 'This guy must be good, he's not wearing a suit!'"

One Last Time

I never wore a suit again for a scientific talk, except in one specific instance.

Many years and lectures later, I was invited to meet with a group of eight major chemical company CEOs. They requested my input on the issue of chlorine toxicity. I was to attend this small, closed-door meeting in Washington with a colleague and friend. He knew me well.

My friend approached me a few days before our flight and said, "Kevin, please wear a suit. If you don't these guys won't be able to hear you. All they'll see is a man not in a suit, thus not deserving of their time or respect."

With some difficulty, he persuaded me to follow his advice, and *boy, was he right*. Those CEOs were all wearing suits that probably cost more than my house. They looked as though they had been born in those suits. They did glance at my crumpled togs with disdain, and then appeared to let it pass. They heard what I had to say, which was important.

That is the only time I can remember wearing a suit for a scientific presentation since that fateful meeting with Jeff in Texas.

Science Survival Tip No. 38: *Dress for your audience, if you want them to hear you.*

POWER OF LANGUAGE AND BELIEF

Science Survival Tip No. 39: *Master at least one foreign language, learn about its culture, and go where they speak it. This will open your mind in ways you cannot imagine.*

I often wondered if my interest in foreign languages and cultures played a role in the way I approach research. Certainly, an interest in learning French took me into the field of toxicology. I had no interest in going from sheep diseases research to killing rats and mice. I just wanted to learn the French language. Which I did, from Proust to Troyat, and *Samedi et à Vous* over a bottle of Feldschlößchen or Cardinal, not that I ever mastered French gender of words. A mouse is female for instance, even a male mouse, while rat is male.

Go figure.

I used to feel bad about my failure to get gender right in French and Spanish, until I recently read a book by Mary Norris. This remarkable linguist admitted that she suffers from the same problem. The book, *Between You And Me, Adventures of a Comma Queen*, is an excellent edification in the art of writing. It's the first book that explained clearly to me when to use "that" *versus* "which" as in the book that/which I liked to read.
Remember, read well to write well.}

A friend, Claire, when she found out what I did at the National Center for Toxicological Research, in Arkansas where we met, named me *Le tueur de petites souris*, or murderer of little mice. But did my studies of French and Spanish, in which I still dream now and then, influence the way I approach scientific research? I suspect they do interact.

Today, I was reading an interesting book on string theory *Not Even Wrong*, by Peter Woit, where it seems that he had had similar thoughts.

The experience of moving from physics to mathematics was somewhat reminiscent of a move in my childhood from the United States to France. Mathematics and physics each have their own distinct and incompatible languages. They often end up discussing the same thing in mutually incomprehensible terms. The differences between the two fields are deeper than simply that of language, involving very distinct histories, cultures, traditions, and modes of thought. Just as in my childhood, I found that there is a lot to learn when one makes such a move, but the extra effort is compensated by an interesting bicultural point of view. I hope to be able to explain some of what I have learned about the complex, continually evolving relationship between the subjects of physics and mathematics and their corresponding academic cultures.

A terrible decision!

I reacted vociferously to a plan by the University of North Carolina at Chapel Hill Medical School to replace a foreign language with a computer language as part of the mandatory graduate school course work. *I was horrified!* That is no substitute for the only road to understanding the cultures of other countries and civilizations. I argued as best I could, but I was overruled.

1975, Edinburgh, Scotland

Having completed my doctorate, I still had a safe job. I was restless, and I wondered what to do next. Then, as usually happens in my case, a crazy idea popped into my head. *I sure would like to speak French better.* I had enjoyed this at school, but why would I think this with three kids and a good job? Although my boyhood home was an emotional train wreck, we were strongly encouraged to gain an education, which included other languages. Why our Mum encouraged this, I do not know, but she did.

I proceeded to seek employment in French-speaking countries all over the world. In fact, we nearly moved to Guelph in Canada. Then I spotted a job in Geneva, Switzerland, as a biochemist/immunologist. I was neither, but both subjects played a role in my doctoral research. I applied and heard nothing from them for six months.

Then the phone rang. Phone calls can change your life. A man, Richard, asked with a strong American accent if Dr. Morgan was on the other end. I replied that I was. He informed me that he was calling from Geneva, Switzerland, and asked if I knew anything about scanning electron microscopy (SEM). I knew a lot about transmission electron microscopy, but I had never even seen an SEM scope, so I responded. *Sure!* I've often wondered why I did that, as I don't habitually tell lies.

Then Richard said, "well, can you come and tell us about it, next week." They had received my application a while ago, where I had mentioned electron microscopy. That job went to someone else, but they were now looking for a pathologist with a background in SEM. Richard said he would put a plane ticket and itinerary in the mail.

I look forward to hearing about your work, Dr. Morgan.

The ticket arrived, and I was firmly on the hook with no SEM data. Geneva sounded like a dream come true, so I guess that is why I responded positively on the spur of the moment. I went into overdrive, phoning around until I found a friendly SEM person, Peter. I explained my dilemma, he laughed and then agreed to help me take an SEM photo of a spore of *Bacillus thiaminolyticus* that I had isolated from a sheep. Peter also kindly explained how to prepare the sample.

I grew some spores, smeared them onto a glass slide, and washed them following Peter's instructions. He then sputter coated them for SEM, and I saw them in 3D, with long ridges that I had only seen in cross section previously. More importantly, I now had an SEM photo linked to my research to take to Switzerland.

I got the job, learned a great deal about *la microscopie électronique à balayage*, and had the time of my life for five years. Our kids ended up bilingual, and the eldest two still are. I even got to carry out my only human study [16], on platelet function in response to aspirin lysinate. There are so many stories from those times, and here is one I love, that has little to do with science.

The jet d'eau in Lake Geneva or Lac Léman, with the Salève in the background.

It was in Switzerland that I experienced the strangest and best dining of my life. It was odd, but effective. I was invited to attend a meal with about 20 colleagues at a famous restaurant in France, located on the other side of the Salève.

The Salève is a long scarp overlooking Geneva. We routinely went out to eat across the border into France, because everything there was about half the price compared to Switzerland because of currency exchange rates at the time. That evening, once through customs, we drove for about an hour while becoming lost in the French countryside before arriving at an old, quaint French farmhouse.

Our entire group arrived almost simultaneously in multiple cars. One of us knocked on the large, ornate, iron-bound, wooden door, to be invited into a warmly-lit foyer by a little old lady. She was 85 to 90 years old, if she was a day! We were politely ushered — or should I say herded — into a large dining room furnished with a beautiful, centrally-placed rectangular, stained and highly polished pine table set for exactly our number. The room had just the right lighting, silver utensils that glittered in response to various candelabra set into the ancient plaster walls that were adorned with old paintings. Carafes of water and bottles of red table wine with fresh French bread in wicker baskets were distributed along the table, all just within reach.

We sat, ate bread as only the French can bake it, and drank some wine, debating a range of topics all the while. *It was just perfect.* As we were starting to wonder when the service was going to start, in comes the same little old lady in a white apron, carrying 20 menus. They were tattered and just right. I do not know why they were just right, they just were. Every aspect of the meal appeared to be carefully orchestrated.

The menus were distributed, comprising a single-folded parchment, which, when opened, revealed two detailed menus, with one on each side. On the left was repas A (meal A) and on the right repas B. The lady then withdrew, and for the next 20 minutes we argued and discussed the meals' various attributes, each having seven or eight courses. And, of course, we consumed more of the excellent bread and wine, but we were now ready for the promised meal to begin. Just as we had each elected either meal A or B, the lady returns with a notepad.

Approaching the person at the head of the table, she enquired as to their chosen meal. This person replied, "I'll have meal B, please, madame." The lady immediately looked apologetic, and stated clearly for all at the table to hear, "I am very sorry, monsieur, but meal B is not available this evening. Are you sure you wouldn't prefer meal A. It is truly delicious."

It was a dance. There was only meal A, and for the entire evening we never did meet anyone but this little old lady, and we heard no noise of cooking or evidence of other staff in the entire house.

Clearly, the French understand the joy of choice as a necessary prelude to dining. We proceeded to have meal A, which I can honestly say was the most enjoyable, best prepared, tastiest and most satisfying dinner of my life. If I remember correctly, it began with liver pate that was perfect [I had yet to embrace my current vegan diet] with even more crusty French baguette, followed by soup, then cheeses (not at the end of the meal), two different meat dishes in rich gravy with assorted vegetables, a small dessert that included some fruit, and finally what I can only describe as an amuse bouche, followed by cognac, and freshly ground coffee as a 'digestif.'

There was no meal B. Only the French could do this with such aplomb, the reasoning for which would be beyond the ken of the crude Anglo-saxon palate.

I loved it!

There is a psychology lesson in there somewhere, but I am not quite sure what it is. Possibly, quality and ambience are as important as quantity, when it comes to food.

All of these wonderful experiences are because I developed a valuable trade and 'bent the truth' of my SEM experience. There is more to a career in science than science, my friends. Most importantly, I discovered that I could have thoughts in French that I could not quite translate correctly into English.

I would like to digress, for a moment, concerning my interest in the intersection of philosophy, religion and science. To me, philosophy is the art of thinking about thinking. My favorite philosophy book is *Sophie's World* by Jostein Gaardner.

The most readable and understandable review of the entire history of philosophy.

It is extremely important to respect others' beliefs. This can be a challenge when employing Darwinian theory of evolution and natural selection. The evening before writing this somewhat challenging chapter, I enjoyed a fascinating discussion with two academics, professors of philosophy and chemistry.

And both are followers of the Bahn'í faith.

Our debate ranged far and wide, from politics to religion, with a fair dose of science. We discussed the challenges of marketing the importance of philosophy to different groups of people with differing beliefs. We considered the challenges of persuading and convincing both others and ourselves of the value of new ideas. We considered the challenges of misoneism (my favorite English word), and how we all resist change, to some degree.

We each described our own personal philosophies, agreeing with Ayn Rand that we all have a philosophy and it's best to have a conscious philosophy as best we can.

My personal philosophy: *I do not know what the hell is going on, but it sure is interesting.*

Then, an image of the following Venn Diagram popped into my head.

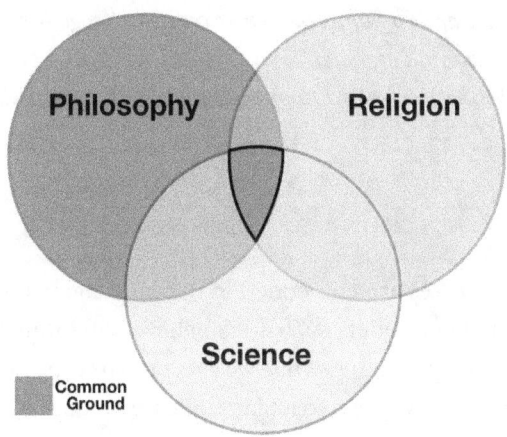

As the debate raged on, I was reminded of the challenges I have recently faced when expressing different scientific and political views online. My struggles were in finding ways to keep people debating issues rather than attacking each other, including yours truly. This is no easy matter.

I suggested that the best way to reach across beliefs, religions, and scientific disciplines, lies in the center of my Venn diagram, where we can start with the common ground and work out from there, to increasingly 'touchy' topics.

Science and the Language of Religion

You might wonder why this issue is so important in the world of science. You might consider religion to have no place in your daily work as a scientist. It certainly does, and it can be a tricky minefield to negotiate.

Dr. John Lightfoot, Vice-Chancellor of the University of Cambridge, and one of the most eminent Hebrew scholars of his time, declared, as the result of his most profound and exhaustive study of the Scriptures, that "heaven and earth, centre and circumference, were created all together, in the same instant, and clouds full of water," and that "this work took place and man was created by the Trinity on October 23, 4004 B.C., at nine o'clock in the morning.

— Andrew D. White

Human cultures are embedded in our languages, and vice versa. This is also true of beliefs. What we believe will influence what we mean when we speak, and what we are prepared to do. I ran into this as a challenge with a particular graduate student.

Yes! I've read it, and I like the ideas of the Jesus character, but the rest is open to debate, in my opinion.

Fall Morning, 2004

I am sitting in a sparsely furnished office at a local university where I had an adjunct faculty position. A meeting has been arranged to set the research course for a graduate student who is sitting opposite me. She had just completed her preliminary examinations after 18 months in the graduate program, and now it was time for her, a mathematics student, to start her post-graduate laboratory work, for which part I had agreed to be her major advisor.

This came as the result of a long-standing collaboration between an applied mathematician at the university and myself. It was an effective way to combine mathematical modeling with laboratory observations and model testing.

I asked her about her interest in mathematics and why she wanted to explore numerical simulation of biological systems. She was coherent, excited about the work, and ready to start. She asked me which project I had in mind for her.

I replied, "Well, I've been collaborating with a group at the US EPA in the RTP, on studies of a rat-adapted influenza virus [80, 100]. My interest is the arms race between the virus and its various hosts. It seems likely that there is a record of this arms race in these viruses and the species they infect. I wonder if we can model the evolution of this arms race."

She immediately responded, quite calmly: "But, I do not believe in evolution. I'm a Christian."

Oh Boy! What can I do about that? Evolutionary theory is one of the bedrocks of biological research at present.

I asked, "So, you think the earth is only 6,000 years old?"

She said she did. In fact, she said she was certain of it.

How about antibiotic resistance?

She said she believed in evolution and natural selection, over the 6,000 years since God had created the Earth.

Now, I had a challenge. This graduate student had spent 18 months finishing her preliminary examinations. She was intelligent, interesting, and I just could not see leaving her high and dry. I thought for a moment, and suggested,

"How about a project that does not depend upon a belief in evolutionary theory. And it is a theory, you know."

She replied that that would be perfect.

"How about helping me to build a crude mathematical model of whole-body energetics or energy metabolism, which we can then employ to explore regulation of hepatic (liver) glycogen regulation. This would provide valuable support to our work on diabetes."

We agreed that this would work and to meet in about a week to discuss the details. In the meantime, I would send a pile of publications for her to read. She stood and headed for the door, but I just could not leave it at that, and said, "If you explore the workings of hepatic glycogen dynamics, won't you be questioning the mind of God, and wouldn't this be blasphemy?"

She smiled and departed. For the next three years this nice student worked hard to save my immortal soul without success. I worked to help her pass her doctoral examination, which I'm pleased to say she did.

All's well that ends well, don't you think?

The language of dreams

I was walking across a rainy parking lot the other day when I espied a rainbow-like pattern on an oil slick. Your first thought might be, *It is just the refraction of light, as in a rainbow, where water droplets act like little prisms.*

You would be wrong.

A delightful little book by a remarkable teacher of science.

These are not refracted colors. They are the consequence of weird interactions between light and matter, as in quantum electrodynamics. I remember thinking, *I'm glad I read that book, QED by Richard Feinman and studied a little math.* Deep down inside, I am a scientist at heart. This reminded me of a dream, and dreams being the language of the subconscious, it seemed that my subconscious often had important things to tell me.

Science Survival Tip No. 40: *Listen to your dreams, they may be wiser than you know.*

I almost moved on down the wrong path, to be saved in the nick of time by a dream.

I was in my mid-60s as a successful and respected (not always liked, which is good) scientist at home mulling over a recent phone call. I had been offered a job. *Not just any job.* Managing a research team of over 200 people. It was flattering to my ego, I guess, but it didn't feel right. Am I at that stage of my career? What's to do?

That night I had a dream, never to be forgotten. My dream-self had been invited out to a special dinner. The meal was in a basement restaurant comprised of a single room with a central table of the highest quality. There were muted lamps around the room with fine art on the walls. The table was set for five, with my empty seat on the right, by the door. Across from my chair, and all the shiny silverwear and cut glass, was a young man. It was my younger competitor, Jack. Not that he was really a competitor more a colleague with similar research interests. Expertise in rat-nose pathology is not a competitive niche, so there were not many of us, mainly Jack, myself, and a few others scattered across the planet.

Science Survival Tip No. 41: *Try to become a big fish in a small pond. Not a pond so small that they might just drain it one day.*

To my left were two unknown people, with the five-member group of dinner guests being completed by a boss figure at the head of the table, to the left farthest from the door. The group was relaxed until the boss figure turned to Jack, saying, "I have an important project that I would like you to take on." That was all he said, but a lot occurred inside my dream-(cathected?)-self.

I felt a flash of anger and resentment toward the boss and towards Jack. *Why didn't I receive this project? Jack is less able than I. Less experienced. It isn't fair.*

This internal rant was rapidly interrupted by Jack, who said, "Kevin, you've got to see this." He passed over a small microscope. I noticed, even while dreaming, that this microscope was the same little monocular that I had purchased as a 13-year old boy. I thought, "That's odd! The lighting of the image is bright and clear, but this microscope only has a mirror to reflect light, and the lighting in this room is too dim. Just candles." I dismissed the thought, and proceeded to see what Jack was talking about, and there it was.

Down the scope was a bacterium, a long helical shape that was much longer than any spirochete I had ever seen, but it had a kink, with one-third of its length at right angles to the rest. I thought, "To twist the body of the organism back into line by pulling the displaced part around would be impossible. The drag coefficient would be too much (I'm a scientist with a fascination for physics even in my dreams, it would appear). I noticed that the bacterium was solving the problem by moving through the disconcerting kink. The short end was becoming shorter as the long end was longer. I briefly thought of DNA, wondering *where is topoisomerase when you need it?* And then I woke up.

The next day, I called to reject the lucrative management position. The dream reminded me that I am a scientist and researcher, not a manager.

I'd be in meetings all day, the horror of horrors.

The Language Of The Subconscious

I have found that Tarot cards can be quite instructive, and I used to read these interesting cards quite frequently. In my opinion, they are not about predicting the future or other "weirdness" (is it really a weird idea?). I suspect that they can help us access the wisdom of our subconscious mind.

About twenty-five years ago, I was asked to set up a task force to address the concerns for chlorine toxicity. This was probably because I'd published some papers on the issue [28, 54, 96, 109]. With some difficulty, I managed to cajole about 15 staff members to consider joining this group. The night before our first planned meeting, just for fun, I decided to do a Tarot reading on the project, using my preferred seven-card Tarot.

An encouraging Tarot reading.

The reading was fascinating in that it made sense of the obstacles that this program would have in its way. More importantly, it presented the issues as beautiful Tarot images rather than a boring PowerPoint list. I welcomed everyone to the meeting, flipped on the projector, and up came my beautiful seven-card Tarot. Imagine each card in the picture has an artistic Tarot drawing, relating to past problems, the present, future, things that will help and hinder, and so forth. All designed to reach the subconscious through myths and metaphors.

The Tarot reading surprised them into a stunned silence at first before they laughed and then they started to understand my logic. This launched our program off on the right note, as they agreed that the reading was surprisingly on target.

Give it a try sometime, you never know. You might have fun with it. Fun is important, even when dealing with toxic gases that save us from cholera.

The Language of Mathematics

I was raised as a visual biologist using images, photos and drawings as my main means of communication. I essentially lived in this picture world until my mid-40s. I understood biology through photos and diagrams or cartoons. That was my world, even though I was aware of mathematics, statistics, and engineering, but my numerical skills were close to zero. I was comfortable in my little visual world. I thought I knew things.

In the early 1980s, I employed a local artist, Andy, to make nasal casts and then hollow molds of the nasal passages of rats.

You can save the animals: One member of our group, Betsy, was always finding ways to save animals, of which we all approved. In order to order rats, you had to go through the song and dance of a study protocol. If your protocol didn't include collecting rat heads for the study of airflow, you couldn't do that. You had to throw the rat heads in the incinerator bucket.

However, Betsy found that it was legal to take rat heads out of this bucket for trial experiments. So, Betsy would find out about a scheduled study by another researcher and ask if she could have the rat heads. They always said yes, as they were going in the incinerator anyway.

When the staff of the study were putting rat heads in the bucket, there would be Betsy's gloved hands waiting to receive them down in the bucket. We did loads of experiments this way, without ever having to kill a rat.

Thanks, Betsy.

What's more, Betsy found out about a scheduled necropsy of a group of monkeys by a different group of researchers. *You guessed it!* We flew to the lab and collected heads that were going to be thrown away, and studied their nasal airflow characteristics [119].

Data without having to kill animals. Who wants to kill animals, anyway. I sure didn't.

It was horrible to see the technicians who had cared for these monkeys for years. They were all crying, especially over one little monkey, called Adolf. I think this was one of the many things that finally contributed to my going vegan.

Thanks again, Betsy.

Construction of the nasal casts and the hollow molds made from them needed someone with real attention to detail. Furthermore, we were using casting techniques that Andy was familiar with. Science moved forward with the help of an artist. We became good friends as a result of working together, and we still are nearly 40 years later. What does this have to do with science?

Everything.

Andy built the bridge in the photo above, on his pond. We would sit on that bridge, have a beer, and talk for ages about our project. Of course, we used Betsy's death-free rat heads for Andy's work. The result of hiring an artist and sculptor was twofold, remarkable nasal casts and molds for our research [72, 101], and a series of shared presentations on art and science. Plus the beer on the bridge, of course.

A few days after one of our talks at Duke, a young scientist, Julie, turns up in my office. She explained that she had attended our talk and was interested in joining our project. I said, "But what use will math be to our project? We use flow tanks and such." *Boy, was I clueless.*

Without saying anything, Julie showed me a diagram on a piece of graph paper, which looked like this:

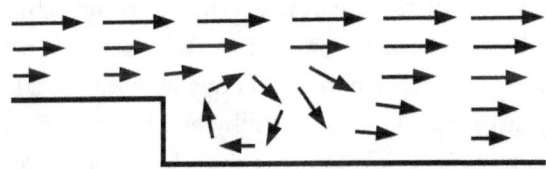

"*That's a backward facing step vortex,*" I said after a glance. I had seen them many times in our flow tank, and in rivers and streams.

But what are those little arrows?

Julii gave me a patient look, replying,

"They're not arrows, they're vectors. The longer the arrow, the faster the flow."

Then I said,

What are vectors and how did you make them? With math?

"I just solved the Navier-Stokes equations in two dimensions for this simple geometry, just to see if I could. My background is in differential geometry, not fluid mechanics, I hasten to add. Do you think I might be able to help with your research program?"

I thought, *Sure as shit, Sherlock, it's like magic.*

I was amazed.

Julie catalyzed my subsequent interest in mathematics. *What a gift that was.* Thanks, Julie.

At the time, I was incredibly ignorant of a technology that had been used to great effect in the aeronautics industry for years:

Computational fluid dynamics. This was the beginning of a fascinating collaboration, and my personal journey into the world of mathematics. It started with my introduction, by Julie, to the Navier-Stokes equations and partial derivatives.

I slowly became obsessed with math, eventually spending five years in my late 50s solving problems for 20 to 30 hours a week. It is the only way to learn math. I worked my way through large quantities of algebra, trigonometry, linear algebra, differential and integral calculus to finally founder on the impenetrable wall of topology.

This was one of the greatest adventures of my aging brain. It started with my meeting Andy, our artist, which then led to my meeting Julie, our mathematician. Funny how the world of science works. My last graduate student, Abby, and last post-doc, Ke, were both mathematicians. We wouldn't even have been able to talk to each other if it wasn't for my late interest in mathematics.

I had many interesting encounters with mathematics during my five years as a kind of math addict. People would come up to me and ask, "What are you doing?" *Studying math for fun.* This would lead to conversations of considerable interest, especially with teenagers. One time, while passing through security at Raleigh Durham International Airport, the two security staff on either side of the X-ray machine asked about my graphing calculator. They were both resentful that they did not have one in school, acting as if it was my fault. All I had in school was a slide-rule and log tables.

Another time, after trying to solve an integral for several weeks for which Mathematica gave a 3-page result (red flag), I was working on it on a plane. A hand reached across, took my mechanical pencil, and a young guy said that he could solve that integral. He worked on it for about half an hour, before saying that it could not be solved because of an error in the problem.

I asked him what he did for a living, Theoretical physicist. They know their math, so I let that problem go.

By the way, in biology you need numerical approximations, as there are rarely analytical solutions. I highly recommend the use of coupled ordinary differential equations to study biological dynamics. You will learn a bunch, especially about what you don't know.

Thirty years later, Julie's mathematical research on airflow in the nose of humans continues, as her work is used to guide nasal surgeons attempting to improve airflow in their patients. And it all started back in 1981 when I looked down the microscope at a section of the nose of a rat exposed to formaldehyde, and said to myself, *I wonder why that is?*

You never know where your questions will carry you.

It's a hard climb to the pChem divide, whichever side you come from.

The pChem Divide

I often ran into trouble when attempting to communicate visual concepts to mathematically trained biologists, especially physical chemists. I also had problems when explaining mathematically-based engineering principles to more visual biologists, notably pathologists.

Yes! I'd evolved into a Jack of all trades, master of one (pathology).

Not everyone enjoys the pChem divide (a term of my own invention). It's much like the continental divide, except it's not water that flows down either side of the mountain range, but people. Young biological researchers seem to flow in one of two directions to more visual or more quantitative approaches. The most quantitative of these people, it seemed the me, are the physical chemists. This is an important discipline in biology, but it's extremely difficult for mathematically challenged biologists. I should know, I am one.

Knowledge on this side of the divide is driven by quantitative simulations, often on computers, and analytical solutions. The denizens of this world will tend to ignore or even scoff at our "pretty pictures." They want numbers. They come largely from the world of physics and mathematics.

We visual biologists draw diagrams while having little to no understanding of the dynamics of the individual components. This all changed for me once I tackled mathematics seriously, in my 50s.

How did I do that? The only way there is. Solve problems. I solved these problems voraciously for about five years, covering the subjects mentioned above. I could generate a simple model, run it in MatLab, and gain a rapid feel for three important aspects of biological research:

- The vast nature of parameter space.
- Dynamics.
- My level of ignorance.

Knowing your level of ignorance is a powerful tool, my friends.

Science Survival Tip No. 42: *Climb the pChcm divide whichever side you are on. You will enjoy great views that reveal fascinating problems for you to solve.*

SCIENCE MANAGEMENT

Science Survival Tip No. 43: *Choose the right currency when negotiating with managers, and take regular psychological vacations or management will drive you crazy.*

Given a chance, science managers will have you in meetings all day so you can't do any science, and then they'll become really mad if you don't do any science, reasonably enough. *You have to escape!*

Meetings are an addictive, highly self-indulgent activity that corporations and other organizations habitually engage in only because they cannot actually masturbate.

— Dave Barry

Meetings can suck the lifeblood right out of you. They proliferate like the pod people in the Body Snatchers. They become self-justifying, but rarely useful. But you can escape, and here is how I learned to do so.

You have to use meetings as currency when you negotiate with managers, as actual science is subservient to "the meeting."

I was walking innocently along a corridor, going from one experiment to another, one paper to another, or just going to get a cup of coffee when I was intercepted by a manager. The manager, lost in the world of meetings, asked if I could come to a meeting Tuesday morning as it was important for me to be there.

Sure it is!!!

I reply that I have plans in the lab at that time, which was true. He looked distressed, saying, "Come on, Kevin. Surely you can move things around and come to our meeting. It's really important."

To managers, all meetings, and in some cases only meetings, are important. They spend so much time in meetings that they come to believe that meetings are where science and other aspects of real life actually happen. It actually happens in people's minds, or in the lab — the lab could be anywhere, by the way, except, as a general rule, in meetings.

I sighed, and said that I would do what I could to reschedule our experiment, which I did. When, a couple of hours later, I called to let him know I would be there, he said, "Great thanks!"

But I could tell that he had moved on. That particular meeting was no longer on his mind — he was thinking about another meeting.

I moved my science around for a meeting, which was probably 99 percent a waste of time, if not 100 percent.

The video, *Meetings Bloody Meetings* starring John Cleese addresses the topic effectively.

Science Survival Tip No. 44: *If you have to run a meeting, keep it as short as possible, have action items for which there is accountability, don't let people grandstand, and get everyone the hell out of there as fast as possible.*

A few months later, I was blithely walking along the same corridor, and the manager comes to me with yet another meeting. "Hey, Kevin," he says, "we have a business-critical meeting on Thursday and we really need you to be there."

A light bulb went on in my head, and I saw an escape (an epiphany), and replied, "I'm sorry, but I'm pretty well tied up with meetings all day Thursday." I had to bend the truth a little, but what are you going to do? Then again, research involves meeting people or equipment, and writing involves meeting with your computer, and so forth.

"Can you reschedule?"

In the blink of an eye, and without any sign of distress, the manager became serious, saying, *"Sure, I understand. I'll see if we can move our meeting to next week. I'll will be back in touch."* He never did get back in touch. Not with respect to that meeting, anyway.

What was the lesson I learned that stood me in such good stead for many years to come and in a wide range of circumstances, not just for science and managers?

If you want to negotiate with someone, step into their world and use their language and values. Think to yourself, *What do they care about?*

You have to use the appropriate currency.

It is all about people, really, which is nicely discussed in *Arrowsmith* by Sinclair Lewis, an insightful book for the aspiring scientist.

Psychological Vacation to Enjoy The Zone

In the mid-90s, I was working with an excellent physicist and engineer, Nick. We were building an oscillating sphere microrheometer, to further my interest in upper respiratory mucociliary function [26, 33, 34, 45, 47, 51]. This device became *my baby*. I spent countless hours, especially during psychological vacations, refining and improving its performance.

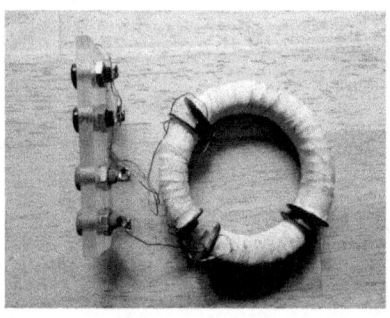

Toroidal magnet, built with the help of professional NASA magnet winders, was an essential component of our oscillating sphere microrheometer.

Psychological vacation, a term of my own invention, is a true vacation from management issues. You spend at least one day per week just doing your own research. Even if you have a large lab, with a lot of staff and students, you must keep your hand in the work of science. Otherwise you will cease to be a scientist, and you will transmogrify into a manager!

Oh No!

One psychological vacation day, a Thursday, I was playing with the rheometer, testing a new idea suggested by a friendly physicist. "Have you considered using white noise instead of stepping through sine wave frequencies one-by-one?"

What a great idea!

Off I went, playing away, watching the oscilloscope and software output, and looking down the microscope at the little (100 micron) nickel ball bouncing back and forth between the poles of the magnet. Could I dissect out storage and loss moduli from these scrambled waveforms, in one fell swoop, I wondered? I bet Fourier can do it!

I started this research at 9:00 a.m. to avoid rush-hour traffic. After what seemed like a few uninterrupted hours, feeling peckish, I took a break for lunch. As I left my lab, I noticed that the institute was oddly silent. On arriving near a window, I found to my complete surprise that it was dark outside.

It was 9:00 p.m.

I had been *in the zone* for 12 hours without a break while completely happy.

That's when you know you are in the right job! You are in your element.

Not when you make the grade in citation index or get some award, that just affects your pay and employment opportunities, not your overall happiness.

A quick note about people who are not in their element in science, of which you need to be aware and beware.

Mimicry and Parasitism

Most populations of organisms suffer from parasites, and the science community is no exception. These are people who take other's ideas and claim them for their own. I have known a few, and tackled a few. If you find them, try to take them down, but they can be slippery devils. Many float to the top of the management pond. I've also known some excellent managers by the way, and it's a skill that escapes me (or bores me to tears, more like).

Mimicry is another tool people use to survive in science when they are unable, or too lazy, to do the work needed. I knew one who was remarkably effective. He would wander around our massive company in a white coat, carrying a clip board. He would do this, whenever he was around, which was infrequent. His behavior was negatively impacting our work, but it still took me a while to get him out of our group.

Science parasites and mimics can be crafty.

Enough said!

MY FINAL LECTURE

Science Survival Tip No. 45: *Every new beginning comes from some other beginning's end.*

— Seneca

Fall 2010

It was the first day of the annual scientific conference of a local toxicology society. A young, up-and-coming scientist in his late 30s, had been selected to chair the first day of a two-day meeting on new technologies. He had invited a Dr. Morgan to be the keynote speaker.

This first speaker at any scientific meeting has a number of jobs to do, including creating excitement around the subject matter. They should briefly lay out the history of the discipline while emphasizing major challenges. Then comes their view of an exciting future for the younger scientists of the day. It is critical that a keynote speaker set the meeting off on an interesting course.

This particular gathering was focused on the challenges of incorporating new techniques into the field of toxicology, or the science of diseases or tissue damage induced by poisonous chemicals or drugs.

A complex subject, rife with opinions and politics, and definitely changing.

―

At that time I was still pretty obsessed with oxidative stress, or redox imbalance. I won't go into the details of our work in this field, but you can find more in the appendix [123, 125]. As we started to work in this area, a friend told me of an expert in the area of oxidative stress biochemistry who worked at Duke, just a few miles away. I am afraid I have forgotten his name, but he was kind enough to meet with me to provide guidance on the topic.

During our meeting, he informed me that he was retiring. He seemed a bit young to retire, probably in his 60s. I asked why, and he replied as follows:

"It's because of quantum mechanics. I just can't come to terms with it. For instance, the methane molecule (CH_3) has two geometric configurations between which it resonates. That's fine, but quantum mechanics says it's in one form or the other, but never in between. I just cannot handle that."

Maybe he should have hung in there. It might not have been true! This interesting and knowledgeable man did help me a great deal with his insights, but we never met again. All because he couldn't handle change, which is what Dr. Morgan's lecture was all about.

———

The young chairman kept glancing toward the door. Everything was ready. Projector, screen, podium, pointer, microphone, audience in their seats, chatting quietly, some sipping coffee. The society members and their guests waited expectantly.

Where was Dr. Morgan?

He had chosen the perfect subject matter for this meeting, *Paradigm Shifts In Biology*.

It's a nerve-racking time running scientific meetings. Plenty can go wrong. Furthermore, a happy audience will speak well of the organization and the performance of the chairperson.

It is already five minutes to eight, the keynote address is scheduled for 8 a.m., and still no sign of the speaker. He has to load his slides in the projector and help the chairman with his brief introduction. The young man was tense, to say the least.

Then the side door opens, and in comes an old guy. He looked athletic, 5 feet 6 inches tall, with a crew cut. Probably in his 60s. He's wearing an Ironman running hat, shorts, and a T-shirt bearing a picture of a cow's skull. *What does he want? Probably a scientist who works in the building, taking a shortcut to his office, or he's one of the maintenance crew. Anyway, he's in the way!*

The old guy approaches the young chairman and introduces himself.

Dr. Morgan?

The old guy seems calm and collected. He requests that they turn off the projector as he has no slides. *No slides for a one-hour time slot?* The audience is already looking their way questioningly. *Only a minute to go!* He steps up to the podium, thanks the chairman politely, and introduces himself and his subject matter.

You have probably guessed that I was that old guy, the keynote speaker of the day. It was my last official scientific lecture as an employee of industry.

I had come a long way since my first scientific lecture. Pathologically shy as a youngster and terrified in 1973 when I gave my first lecture on a nervous system disease in sheep. During the intervening years, I had learned that lecturing is show business. Four years in Toast Masters, followed by hundreds of lectures with both positive and negative feedback from my audiences and students had honed my skills.

I sure knew plenty about the changes that had occurred in science, especially biological research, over the previous 55 years. Computers provided the power engine for many of the big changes in the way that we investigated and thought about biological systems and chemically induced diseases. Computers opened the road to *omics*, but I am getting ahead of myself.

As I stood on that podium in 2010, I saw an excited group of young scientists. The majority were 45 years old or younger. This caused me to reflect on my early years as a young scientist. All that history flashing through my mind. Powerful images around which to build my stories for the next hour. The power and speed of these mental images makes me think that interference pattern-based holonomic brain theory makes total sense.

I used to scratch my head about the mechanisms underlying the instant recall of images in our heads, but now I am free of that conundrum. I find that some questions turn into a constant itch.

Standing at the podium my presentation followed my life as a scientist, some of which has been covered in this book. There are tricks to surviving in science, many of which are covered in this excellent book by Carl J. Sindermann. Highly recommended.

When I was in my 40s, an older guy who was at least 70 (so old), gave me some valuable advice, which I will pass along to you now.

Kevin, things change. We age. There is always a right time to change. Soon you'll be in your 50s and 60s. That will be the time to stop building your career, and to focus on helping younger researchers to replace you. It's healthy. It's normal. You can practice now, by going to a poster session to find a young student standing alone by their beloved poster. You might not be particularly interested in the subject matter, but young people need encouragement. Go to such a poster, read it, and engage the young person in a conversation about their work. Remember, it's not about you, it's not about them, it's about our love of science, which will continue when we're gone.

From then on, and for the next 20 years, I followed his advice at the many scientific meetings I attended. Each time I was rewarded by a smile and an obvious sense of relief that someone was interested in their work.

I call this process, "The Gaussian Life."

The Wonderful Journey Called Life

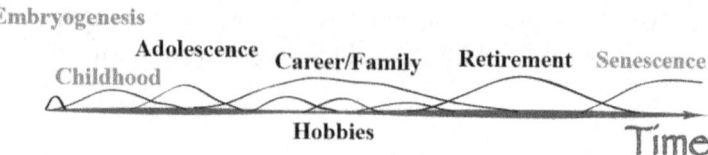

During my last lecture, I decided to leave science to set up a business. Several young members of the audience came up to me afterwards, with one asking if I was retiring. It sounded as though you were saying *sayonara*. I replied that 40 years was enough. It was time to move on. It was their turn to be at the podium, building science careers. It is the only way science can move forward.

Science Survival Tip No. 46: *A new scientific truth does not triumph by convincing its opponents and making them see the light, but rather its opponents eventually die, and a new generation grows up that is familiar with it.*

— Max Planck

Finally, one of these young scientists said, "What would you do if you were us, starting our career in the biological sciences?"
I replied, without hesitation.
I would choose a trade, master it, then study mathematics and multiple branches of engineering, as cells are exquisite engineers. It is what they do!

THE SCIENCE SURVIVAL TIPS

Science Survival Tip No. 1: *It is not the strongest of the species that survives, nor the most intelligent. It is the one that is most adaptable to change.* – Charles Darwin

Science Survival Tip No. 2: *Be a Jack of all trades, master of ONE.*

Science Survival Tip No. 3: *Don't panic, and remember to bring a large bath towel and a SEP field generator.*

Science Survival Tip No. 4: *Never give up trying to fly. I don't care how old you are.*

Science Survival Tip No. 5: *Learn how to question the obvious, which is almost never obvious.*

Science Survival Tip No. 6: *The key currency of science, most of the time, is publications in peer reviewed scientific journals, so publish, publish, publish.*

Science Survival Tip No. 7: *Observe and question the obvious before attempting to solve.*

Science Survival Tip No. 8: *You only see what you know, so learn and learn some more, and you might see more than you can ever imagine.*

Science Survival Tip No. 9: *Experience is the best teacher (but only when the experience isn't fatal).* — Peter J Feibelman

Science Survival Tip No. 10: *Ideas come from other ideas, and to be seen as a genius, rather than a pirate, hide your sources.* – David Kord Murray and Albert Einstein

Science Survival Tip No. 11: *The answers you get depend upon the questions you ask.* — Thomas Kuhn

Science Survival Tip No. 12: *Do not let everyone in your lab, especially theoreticians, as they might break your equipment. I'm not joking.*

Science Survival Tip No. 13: *Do not reinvent the wheel if you can copy someone else's design, but remember to give them credit.*

Science Survival Tip No. 14: *Do not publicly say "I told you so," it achieves nothing useful.*

Science Survival Tip No. 15: *If I had an hour to solve a problem, I'd spend 55 minutes thinking about the problem, and five minutes thinking about solutions.* — Albert Einstein

Science Survival Tip No. 16: *It's OK to stand on the shoulders of giants whenever you get a chance. The view is better from up there.*

Science Survival Tip No. 17: *Any effort that has self glorification as its final endpoint is bound to end in disaster.*— Robert M. Pirsig

Science Survival Tip No. 18: *In science, it does not matter who is right, it only matters what is right.*

Science Survival Tip No. 19: *In conflict, be fair and generous.* — Lao-tzu

Science Survival Tip No. 20: *Be good to people on the way up, you may meet them again on the way down.*

Science Survival Tip No. 21: *Do not let stodgy people put down your inner child, to turn creative play into tedious work.*

Science Survival Tip No. 22: *Find out who you are and do it on purpose.* — Dolly Parton

Science Survival Tips No. 23: *Cut those unhealthy childhood tapes in your head, for real freedom of thought and personal happiness.*

Science Survival Tip No. 24: *Listen to your body, because it will let you know if you're in your optimal job.*

Science Survival Tip No. 25: *For most of us the problem isn't that we aim too high and fail - it's just the opposite - we aim too low and succeed.* — Ken Robinson

Science Survival Tip No. 26: *Be unreasonable - The reasonable man [person] adapts himself to the world; the unreasonable one persists in trying to adapt the world to himself. Therefore, all progress depends on the unreasonable man [person].* — George Bernard Shaw

Science Survival Tip No. 27: *If people get in your way, beat them at their own game, to earn their respect or at least to make them fear you.*

Science Survival Tip No. 28: *It ain't what you don't know that gets you into trouble. It's what you know for sure that just ain't so.* — Mark Twain

Science Survival Tip No. 29: *Scientific disciplines are human only tools. They come and go - when did you last consult your alchemist or necromancer, for instance?*

Science Survival Tip No. 30: *Read well, to write well.*

Survival Tip No. 31: *Consider playing with the old methods, or at least read of their history.* For instance, puzzled by statistics? Read *The Lady Tasting Tea*, by David Salsburg.

Science Survival Tip No. 32: *Don't worry about people stealing an idea. If it's original, you will have to ram it down their throats.* — Howard H. Aiken

Science Survival Tip No. 33: *There's always a way to overcome roadblocks, you just have to find them.*

Science Survival Tip No. 34: *Adapt what is useful, reject what is useless, and add what is specifically your own.* — Bruce Lee

Science Survival Tip No. 35: *Just because you love your data, do not expect the world to be as equally impressed. However, one discovery can change the world, even if you are not there to see it change.*

Science Survival Tip No. 36: *Those who tell (and live) the best stories will rule the future.* — Jonah Sachs

Science Survival Tip No. 37: *It is alright to wear the same clothes everyday, remember to bite the hand that feeds you, and never put on the golden handcuffs.*

Science Survival Tip No. 38: *Dress for your audience, if you want them to hear you.*

Science Survival Tip No. 39: *Master at least one foreign*

language, learn about its culture, and go where they speak it. This will open your mind in ways you cannot imagine.

Science Survival Tip No. 40: *Listen to your dreams, they may be wiser than you know.*

Science Survival Tip No. 41: *Try to become a big fish in a small pond. Not a pond so small that they might just drain it one day.*

Science Survival Tip No. 42: *Climb the pChcm divide whichever side you are on. You will enjoy great views that reveal fascinating problems for you to solve.*

Science Survival Tip No. 43: *Choose the right currency when negotiating with managers, and take regular psychological vacations or management will drive you crazy.*

Science Survival Tip No. 44: *If you have to run a meeting, keep it as short as possible, have action items for which there is accountability, don't let people grandstand, and get everyone the hell out of there as fast as possible.*

Science Survival Tip No. 45: *Every new beginning comes from some other beginning's end.* — Seneca

Science Survival Tip No. 46: *A new scientific truth does not triumph by convincing its opponents and making them see the light, but rather its opponents eventually die, and a new generation grows up that is familiar with it.* — Max Planck

ACKNOWLEDGMENTS

I couldn't have done this on my own, so thank you everyone, including all the people mentioned in the narrative in addition to the many 100s of you who are not.

APPENDIX: THE AUTHOR'S PEER REVIEWED JOURNAL ARTICLES

1. Morgan, K.T. (1973). Chronic copper toxicity of sheep: an ultrastructural study of spongiform leucoencephalopathy. Res. Vet. Sci., 15, 88-95.
2. Morgan, K.T. (1974). An ultrastructural study of ovine polioencephalomalacia. J. Pathol., 110, 123-130.
3. Morgan, K.T. (1974). Amprolium poisoning of preruminant lambs: an ultrastructural study of the cerebral malacia and the nature of the inflammatory response. J. Pathol., 112, 229-236.
4. Morgan, K.T. and Kelly, B.G. (1974). Ultrastructural study of brain lesions produced in mice by the administration of Clostridium welchii Type D Toxin. J. Comp. Pathol., 84, 181-191.
5. Angus, K.W., Sykes, A.R., Gardiner, A.C. and Morgan, K.T. (1974). Mesangiocapillary glomerulonephritis in lambs. 1. clinical and biochemical findings in a Finnish Landrace Flock. J. Comp. Pathol., 84, 309-307.
6. Angus, K.W., Gardiner, A.C., Morgan, K.T. and Gray, E.W. (1974). Mesangiocapillary glomerulonephritis in lambs. 2. pathological findings and electron microscopy of the renal lesions. J. Comp. Pathol., 84, 319-330.
7. Morgan, K.T. and Lawson, G.H.K. (1974). Thiaminase type I

producing bacilli and ovine polioencephalomacia. Vet. Record., 96, 361-363.

8. Morgan, K.T., Coop, R.L. and Doxey, D. (1975). Amprolium poisoning of preruminant lambs: an investigation of the encephalopathy and the haemorrhagic and diarrhoeic syndromes. J. Pathol., 116, 73-81.

9. Morgan, K.T., Kelly, B.G. and Buxton, D. (1975). Vascular leakage produced in the brains of mice by Clostridium welchii type D toxin. J. Comp. Pathol., 85, 461-466.

10. Buxton, D. and Morgan, K.T. (1976). Studies of lesions produced in the brains of mice by Clostridium welchii type D toxin. J. Comp. Pathol., 86, 435-447.

11. Morgan, K.T. and Lawson, G.H.K. (1977). The inhibition of thiaminase type I-producing bacteria by other bacterial species and volatile fatty acids. J. Appl. Bacteriol., 42, 311-319.

12. Morgan, K.T., Gardiner, A.C. and Angus, K.W. (1977). Studies of lesions in the choroid plexus of lambs affected with an acute spontaneous mesangiocapillary glomerulonephritis. J. Comp. Pathol., 87, 15-27.

13. Linklater, K., Dysons, D. and Morgan, K.T. (1977). Investigations of outbreaks of ovine polioencephalomalacia. Res. Vet. Sci., 22, 308-312.

14. Morgan, K.T. and Strolin-Benedetti, M. (1977). Localization of a new neuroleptic in the pituitary gland of the rat. Experientia, 33, 1485-1486.

15. Strolin-Benedetti, M., Donath, A., Frigerio, A., Morgan, K.T., Laville, L., Malnoe, A. and Larue, D.A. (1978). Absorption, Elimination et metabolisme d'un nouveau medicament neuroleptique (FLO 1374) Chez le rat, le chien et l'homme. Ann. Pharm. Francaises, 36, 279-288.

16. Morgan, K.T., Duchosal, F., Rogg, C. and Miescher, P.A. (1980). Effect of aspirin and aspirin lysinate on platelet function in smokers and non-smokers. Acta Haematol., 63, 177-184.

17. Coggins, C.R.E., Fouillet, X.L.M., Lam, R. and Morgan, K.T. (1980). Cigarette smoke induced pathology of the rat respira-

tory tract: a comparison of the effects of the particulate and vapour phases. Toxicology, 16, 83-101.

18. Morgan, K.T., Crowder, D.M., Frith, C.H., Littlefield, N. and Coleman. (1981). Spongiform leukoencephalopathy induced in mice by oral benzidine administration. Toxicol. Pathol., 9, 4-8.

19. Frith, C.H., Zuna, R.E. and Morgan, K.T. (1981). A morphologic classification and incidence of spontaneous ovarian neoplasms in three strains of mice. J. Natl. Cancer Inst., 67, 693-702.

20. Musy, C., Morgan, K.T. and Coggins, C.R.E. (1981). Computerization of data capture for mouse skin painting studies. Int. J. Biomed. Computing. 12, 419-431.

21. Coggins, C.R.E., Lam, R. and Morgan, K.T. (1982). Chronic inhalation study in rats, using cigarettes containing different amounts of cytrel tobacco supplement. Toxicology, 22, 287-296.

22. Coggins, C.R.E., Haroz, R.K., Lam, R. and Morgan, K.T. The tumourigenicity of smoke condensates from cigarettes containing different amounts of cytrel, as assessed by mouse skin painting. Toxicology, 23, 177-185.

23 Morgan, K.T., Swenberg, J.A., Hamm, T.E.Jr., Wolkowski-Tyl, R. and Phelps, M. (1982). Histopathology of acute toxic response in rats and mice exposed to methyl chloride by inhalation. Fundam. Appl. Toxicol., 2, 293-299.

24. Morgan, K.T., Johnson, B.P., Frith, C.H. and Townsend, J. (1982). An ultrastructural study of spontaneous mineralization in the brains of aging mice. Acta Neuropathol., 58, 120-124.

25. Chapin, R.E., Morgan, K.T. and Bus, J.S. (1983). The morphogenesis of testicular degeneration induced in rats by orally administered 2,5-hexanedione. Exp. Mol. Pathol., 38, 149-169.

26. Morgan, K.T., Patterson, D.L. and Gross, E.A. (1984). Frog palate mucociliary apparatus: structure, function and response to formaldehyde gas. Fundam. Appl. Toxicol., 4, 58-68.

27. Morgan, K.T., Frith, C.H., Swenberg, J.A., McGrath, J.T., Zulch, K.J. and Crowder, D.M. (1984). A morphologic classification

of brain tumors found in several strains of mice. J. Natl. Cancer Inst., 72, 151-160.

28. Jiang, X.Z., Buckley, L.A. and Morgan, K.T. (1983). Histopathology of toxic responses to chlorine gas in the nasal passages of rats and mice. Toxicol. Appl. Pharmacol., 71, 225-236.

29. Swenberg, J.A., Barrow, C.S., Morgan, K.T., Heck, H.d'A, Boreiko, C.J., Levine, R.J. and Starr, T.B. (1983). Non-linear biological responses to formaldehyde and their implications for carcinogenic risk assessment. Carcinogenesis. 4, 945-952.

30. Greenman, D.L., Highman, B., Kodell, R., Morgan, K.T. and Norvell, M. (1984). Neoplastic and non-neoplastic responses to diethylstilbestrol in C3H mice. J. Toxicol. Environ. Health., 14, 551-567.

31. Buckley, L.A., Jiang, X.Z., James, R.A., Morgan, K.T. and Barrow, C.S. (1984). Respiratory tract lesions induced by sensory irritants at the RD50 concentration. Toxicol. Appl. Pharmacol., 74, 417-429.

32. Chapin, R.E., White, R.D., Morgan, K.T. and Bus, J.S. (1984). Studies of lesions induced in the testis and epididymis of F344 rats by inhaled methyl chloride. Toxicol. Appl. Pharmacol., 76, 328-343.

33. Morgan, K.T., Jiang, X.Z., Patterson, D.L. and Gross, E.A. (1984). The nasal mucociliary apparatus: correlation of structure and function in the rat. Am. Rev. Resp. Dis., 130, 275-281.

34. Morgan, K.T., Jiang, X.Z., Gross, E.A. and Patterson, D.L. (1985). A procedure for the study of effects of irritant gases on the nasal mucociliary apparatus of rats. Acta Pharmacol. Sinica. 6, 111-114. (Chinese)

35. Jiang, X.Z., White, R.D and Morgan, K.T. (1985). An ultrastructural study of lesions induced in the cerebellum of mice by inhalation exposure to methyl chloride. Neurotoxicol., 6, 93-104.

36. Morgan, K.T., Gross, E.A., Lyght, O. and Bond, J.A. (1985). Morphologic and biochemical studies of a nitrobenzene-induced encephalopathy in rats. Neurotoxicol., 6, 105-116.

37. Buckley, L.A., Morgan, K.T., Swenberg, J.A., James, R.A.,

Hamm, T.E.Jr., and Barrow, C.S. (1985). The toxicity of dimethylamine in F344 rats and B6C3F1 mice following a 1 year inhalation exposure. Fundam. Appl. Toxicol., 5, 341-352.

38. Beauchamp, R.O., Kligerman, A.D., Andjelkovich, D.A., Morgan, K.T., and Heck, H.d'A. (1985). A critical review of the literature on acrolein toxicity. C.R.C. Critical Reviews in Toxicology, 14, 309-380.

39. Welsch, F. and Morgan, K.T. (1985). Placental transfer and developmental toxicity of 2,2',4,4',5,5'-hexabromobiphenyl in B6C3F1 mice. Toxicol. Appl. Pharmacol., 81, 431-442.

40. Morgan, K.T., Jiang, X.Z., Starr, T.B. and Kerns, W.D. (1986). More precise localization of nasal tumors associated with chronic exposure of F-344 rats to formaldehyde gas. Toxicol. Appl. Pharmacol., 82, 264-271.

41. Morgan, K.T., Patterson, D.L. and Gross, E.A. (1986). Responses of the nasal mucociliary apparatus of F344 rats to formaldehyde gas. Toxicol. Appl. Pharmacol., 82, 1-13.

42. Bogdanffy, M., Randall, H.W. and Morgan, K.T. (1985). Histochemical localization of aldehyde dehydrogenase activity in the nasal passages of F-344 Rats. Toxicol. Appl. Pharmacol., 82, 560-567.

43. Chellman, G.J., Morgan, K.T., Bus, J.S. and Working, P.K. (1986). Inhibition of methyl chloride toxicity in male F-344 rats by the anti-inflammatory agent BW755C. Toxicol. Appl. Pharmacol., 85, 367-379.

44. Bogdanffy, M.S., Randall, H.W. and Morgan, K.T. (1986). Biochemical quantitation and histochemical localization of carboxylesterase in the nasal passages of the Fischer-344 rat and B6C3F1 mouse. Toxicol. Appl. Pharmacol., 88, 183-194.

45. Morgan, K.T., Gross, E.A. and Patterson, D.L. (1986). Distribution, progression, and recovery of acute formaldehyde-induced inhibition of nasal mucociliary function in F-344 rats. Toxicol. Appl. Pharmacol., 86, 448-456.

46. Randall, H.W., Bogdanffy, M.S. and Morgan, K.T. (1987).

Enzyme histochemistry of the rat nasal mucosa embedded in cold glycol methacrylate (GMA). Am. J. Anat., 179, 10-17.

47. Bogdanffy, M.S., Morgan, P.H., Morgan, K.T. and Starr, T.B. (1986). Binding of formaldehyde to human and rat nasal mucus and bovine serum albumin. Toxicol. Letters. 38, 145-154.

48. Patra, A.L., Gooya, A. and Morgan, K.T. (1986). Airflow characteristics in a baboon nasal passage cast. J. Appl. Physiol., 61, 1959-1966.

49. Greene, J.A., Sleet, R.B., Morgan, K.T. and Welsch, F. (1986). Cytotoxic effects of ethylene glycol monomethyl ether in the forelimb bud of the mouse embryo. Teratology, 36, 23-34.

50. Alison, R.H., Morgan, K.T. and Haseman, J.K. (1986). Morphology and classification of ovarian neoplasms in F-344 rats and B6C3F1 mice. J. Natl. Cancer Inst., 78, 1229-1243.

51. Gross, E.A., Patterson, D.L. and Morgan, K.T. (1987). Effects of acute and chronic dimethylamine exposure on the nasal mucociliary apparatus. of F-344 rats. Toxicol. Appl. Pharmacol., 90, 359-376.

52. Hurtt, M.E., Morgan, K.T., and Working, P.K. (1987). Histopathology of acute toxic responses in selected tissues from rats exposed by inhalation to methyl bromide. Fundam. Appl. Toxicol., 9, 352-365.

53. Hurtt, M.E., Thomas. D.A., Working, P.K., Monticello, T.M., and Morgan, K.T. (1988). Degeneration and regeneration of the olfactory epithelium following inhalation exposure to methyl bromide: pathology, cell kinetics, and olfactory function. Toxicol. Appl. Pharmacol., 94, 311-328.

54. Klonne, D.R., Ulrich, C.E., Riley, M.G., Hamm, T.E.Jr., Morgan, K.T., and Barrow, C.S. (1987). One-year inhalation toxicity study of chlorine in rhesus monkeys (Macaca mulatta). Fundam. Appl. Toxicol., 9, 557-572.

55. Monticello, T.M., Morgan, K.T., Everitt, J.I., and Popp, J.A. (1989). Effects of formaldehyde gas on the respiratory tract of rhesus monkeys: pathology and cell proliferation. Am. J. Pathol., 134, 515-527.

56. Peele, D.B., Alison, S.D., Bolon, B., Prah, J.D., Jensen, K.F., and Morgan, K.T. (1990). Functional deficits produced by 3-methylindole-induced olfactory mucosal damage revealed by a simple olfactory learning task. Toxicol. Appl. Pharmacol, 107, 191-202.

57. Morgan, K.T. and Monticello, T.M. (1990). Airflow, gas deposition, and lesion distribution in the nasal passages. Environ. Health Perspect., 85, 209-218.

58. Monticello, T.M., Morgan, K.T. and Uriah, L. (1990). Non-neoplastic nasal lesions in rats and mice. Environ. Health Perspect., 85, 249-274.

59. Randall, H.W., Monticello, T.M., and Morgan, K.T. (1989). Large area sectioning for morphologic studies of nonhuman primate nasal cavities. Stain Technol., 63, 355-362.

60. Davenport, C.J., and Morgan, K.T. (1989). In vitro neurotoxicology: industrial applications. In Vitro Toxicology, 2, 207-208.

61. Monticello, T.M., Morgan, K.T. and Hurtt, M. (1990). Unit length as the denominator for quantitation of cell proliferation rate in nasal epithelium. Toxicol. Pathol., 18, 24-31.

62. Goldsworthy, T.L., Monticello, T.M., Morgan, K.T., Bermudez, E., Wilson, D., Jackh, R., and Butterworth, B.E. (1991). Examination of potential mechanisms of carcinogenicity of 1,4-dioxane in rat nasal epithelial cells and hepatocytes. Arch. Toxicol., 65, 1-9.

63. Keller, D., Randall, H.W., Heck, H.d'A, and Morgan, K.T. (1990). Histochemical localization of formaldehyde dehydrogenase. Toxicol. Appl. Pharmacol., 106, 311-326.

64. Bonnefoi, M., Davenport, C., and Morgan, K.T. (1990). Metabolism and toxicity of methyl iodide in primary dissociated neural cell cultures. Neurotoxicol., 12, 33-46.

65. Davenport, C., Bonnefoi, M., Williams, D., and Morgan, K.T. (1992). In vitro neurotoxicity of methyl iodide. Toxicol. in Vitro, 6, 11-20.

67. St. Clair, M.B., Gross, E.A., and Morgan, K.T. (1990).

Comparative nasal toxicity of formaldehyde and glutaraldehyde in F-344 rats using nasal instillation. Toxicol. Pathol., 18, 353-361.

68. Cheng, Y.S., Weh, L., Hansen, B., and Morgan, K.T. (1990). Deposition of ultrafine aerosols in the nasal casts of F-344 rats. Toxicol. Appl. Pharmacol, 106, 222-233.

69. Bonnefoi, M., Monticello, T.M., and Morgan, K.T. (199?). Toxic and neoplastic responses in the nasal passages: future research needs. Exp. Lung Res., 17, 853-868.

70. Casanova, M., Morgan, K.T., Steinhagen, W.H., Everitt, J.I., Popp, J.A. and Heck, H.d'A. (1991). Covalent binding of inhaled formaldehyde to DNA in the respiratory tract of rhesus monkeys: pharmacokinetics, rat-to-monkey interspecies scaling, and extrapolation to man. Fundam. Appl. Toxicol., 17, 409-428.

71. Kimbell, J.S. and Morgan, K.T. (1991). Airflow effects on regional disposition of particles and gases in the upper respiratory tract. Rad. Dos. Prot., 38, 213-219.

72. Morgan, K.T., Kimbell, J.S., Monticello, T.M., Patra, A.L. and Fleishman, A. (1991). Studies of inspiratory airflow patterns in the nasal passages of the F344 rat and rhesus monkey using nasal molds: relevance to formaldehyde toxicity. Toxicol. Appl. Pharmacol, 110, 223-240.

73. Monticello, T.M., Miller, F.J., and Morgan, K.T. (1991). Regional increases in rat nasal respiratory epithelial cell proliferation following acute and subacute inhalation of formaldehyde. Toxicol. Appl. Pharmacol., 111-409-421.

74. Davenport, C., Ali, S., Miller, F.J., Lipe, G., Morgan, K.T., and Bonnefoi, M. (1991). Effect of methyl bromide on regional brain glutathione, glutathione-S-transferases, monoamines, and amino acids in F344 rats. Toxicol. Appl. Pharmacol., 112-120-127.

75. Bolon, B., Bonnefoi, M.S., Roberts, K.C., Marshall, M.W., and Morgan, K.T. (1991). Toxic interactions in the rat nose: pollutants from soiled bedding and methyl bromide. Toxicol. Pathol., 19, 571-579.

76. Morgan, K.T. (1991). Approaches to the identification and

recording of nasal lesions in toxicology studies. Toxicol. Pathol., 19, 337-351.

77. Brown, H.R., Monticello, T.M., Maronpot, R.R., Randall, H.W., Hotchkiss, J.R., and Morgan, K.T. (1991). Proliferative and neoplastic lesions in the rodent nasal cavity. Toxicol. Pathol., 19, 358-372.

78. Genter, M.B., Llorens, J., O'Callaghan, J.P., Peele, D.B., Morgan, K.T., and Crofton, K.M. (1992). Olfactory toxicity of ß,ß'-Iminodiproprionitrile (IDPN) in the rat. J. Pharm. Exp. Ther., 263, 1432-1439.

79. Dasgupta, A., Guenard, P., Ultman, J.S., Kimbell, J.S., and Morgan, K.T. (1992). A photographic method for the visualization of mass uptake patterns in aqueous systems. Int. J. Heat Mass Transfer, 36, 453-462.

80. Dye, J.A., Morgan, K.T., Neldon, D.L., Tepper, J.S., Burleson, G.R. and Costa, D.L. (1994). Characterization of respiratory disease in rats following neonatal inoculation with a rat-adapted influenza virus. Vet. Pathol., 33, 43-54.

81. Kimbell, J.S., Gross, E.A., Joyner, D.R., Godo, M.N. and Morgan, K.T. (1993). Application of computational fluid dynamics to regional dosimetry of inhaled chemicals in the upper respiratory tract of the rat. Toxicol. Appl. Pharmacol., 121, 253-263.

82. Conolly, R.B., Morgan, K.T., Andersen, M.E. Monticello, T.M., and Clewell, H.J. (1992). A biologically-based risk assessment strategy for inhaled formaldehyde. Comments on Toxicology, 4, 269-293.

83. Beauchamp, R.O. Jr., St. Clair, M.B., Fennell, T.R., Clarke, D.O., Kari, F.W., and Morgan, K.T. (1992). Critical review - toxicology of glutaraldehyde. Crit. Rev. Toxicol., 22, 143-174.

84. Bolon, B., Dorman, D.C., Janszen, D., Morgan, K.T., and Welsch, F. (1993). Phase-specific developmental toxicity in mice following maternal methanol inhalation. Fundam. Appl. Toxicol., 21, 508-516

85. Bermudez, E., Chen, Z., Gross, E.A., Walker, C.L., Recio, L., Sisk, S., Pluta, L., and Morgan, K.T. (1994). Characterization of

cell lines from formaldehyde-induced rat nasal squamous cell carcinomas. Molecular Carcinogenesis, 9, 193-199.

86. Recio, L., Sisk, S., Pluta, L., Bermudez, E., Gross, E.A., Chen, Z., Morgan, K.T., and Walker, C. (1992). p53 Mutations in formaldehyde-induced nasal squamous cell carcinomas in rats. Cancer Res., 52, 6113-6116.

87. Dasgupta, A., Guenard, P., Anderson, S.M., Ultman, J.S., and Morgan, K.T. (1992). Calibration of a photographic method for imaging mass transfer in aqueous solutions. Int. J. Heat Mass Transfer, 38, 2029-2037.

88. Bolon, B., Dorman, D.C., Bonnefoi, M.S., Randall, H.W., and Morgan, K.T. (1993). Histopathologic approaches to chemical toxicity using primary cultures of dissociated neural cells grown in chamber slides. Toxicol. Pathol., 21, 465-479.

89. Dorman, D.C., Bolon, B., and Morgan, K.T. (1993). The toxic effects of formate in dissociated primary mouse neural cell cultures. Toxicol. Appl. Pharmacol., 122, 265-272.

90. Bolon, B., Welsch, F., and Morgan, K.T. (1994). Methanol-induced neural tube defects in mice: pathogenesis during neurulation. Teratology, 49, 497-517.

91. Larson, J.L., Wolf, D.C., Morgan, K.T., Mery, S., and Butterworth, B.E. (1994). The toxicity of one week exposures to inhaled chloroform in female B6C3F1 mice and male F344 rats. Fundam. Appl. Pharmacol., In Press.

92. Mery, S., Larson, J.L., Butterworth, B.E., Wolf, D.C., Harden, R., and Morgan, K.T. (1994). Nasal toxicity of chloroform in male F344 rats and female B6C3F1 mice following 1-week inhalation exposure. Toxicol. Appl. Pharmacol., 125, 214-227.

93. Gross, E.A., Mellick, P.W., Kari, F.W., Miller, F.J., and Morgan, K.T. (1994). Histopathology and cell replication responses in the respiratory tract of rats and mice exposed by inhalation to glutaraldehyde for up to thirteen weeks. Fundam. Appl. Toxicol., 23, 348-362.

94. Douglas C. Wolf, Elizabeth A. Gross, Otis Lyght, Edilberto Bermudez, Leslie Recio, and Kevin T. Morgan (1995). Immunohis-

tochemical localization of p53, PCNA and TGF-alpha proteins in formaldehyde-induced rat nasal squamous cell carcinomas. Toxicol. Appl. Pharmacol., 132, 27-35.

95. Mery, S., Gross, E.A., Joyner, D.R., Godo, M., and Morgan, K.T. (1994). Nasal diagrams: a tool for recording the distribution of nasal lesions in rats and mice. Toxicol. Pathol., 22, 353-372.

96. Wolf, D.C., Morgan, K.T., Gross, E.A., Barrow, C., Moss, O., James, R.A., and Popp, J.A. (1995). Two-year inhalation exposure of female and male B6C3F1 and F344 rats to chlorine gas induces lesions confined to the nose. Fundam. Appl. Toxicol., 24, 111-131.

97. Casanova, M., Morgan, K.T., Gross, E.A., Moss, O.R., and Heck, H.d'A (1994). DNA-protein cross-links and cell replication at specific sites in the nose of F344 rats exposed subchronically to formaldehyde. Fundam. Appl. Toxicol., 23, 525-536.

98. Godo, M.N., Richardson, R.B., Morgan, K.T. and Kimbell, J.S. (1995). Reconstruction of complex passageways for simulations of transport phenomena: development of a graphical user interface for biological applications. Computer Methods and Programs in Biomedicine, 47, 97-112.

99. Larson, J. , Wolf, D., Mery, S., Morgan, K.T. and Butterworth, B.E. (1995). Toxicity and cell proliferation in the liver, kidneys and nasal passages of female F-344 rats, induced by chloroform administered by gavage. Fd. Chem. Toxic., 33, 443-456.

100. Ibanes, J.D., Morgan, K.T., and Burleson, G.R. (1996). Histopathological changes in the upper and lower respiratory tract of F344 rats following infection with a rat adapted influenza virus. Vet. Pathol., 33, 412-418.

101. Kepler, G.M., Joyner, D.R., Fleishman, A., Richardson, R., Gross, E.A., Morgan, K.T., Kimbell, J.S, and Godo, M.N. (1995). Method for obtaining accurate geometrical coordinates of nasal airways for computer dosimetry modeling and lesion mapping. Inhalation Toxicol., 7, 1207-1224.

102. Larson, J.L., Templin, M.V., Wolf, K.C., Jamison, J.R., Leininger, J.R., Mery, S., Morgan, K.T., Wong, B.A., Conolly, R.B.

and Butterworth, B.E. (1995). A 90-day chloroform inhalation study in female and male B6C3F1 mice: implications for cancer risk assessment. Fundam. Appl. Toxicol., 30, 118-137.

103. Butterworth, B.E., Conolly, R.B., and Morgan, K.T. (1995). A strategy for establishing mode of action of chemical carcinogens as a guide for approaches to risk assessments. Cancer Letters, 93, 129-146.

104. Dorman, D.C., Struve, M.F., Ritzman, T.K., Owens, J.G. and Morgan, K.T. (1996). Acrylonitrile-induced cytotoxicity in primary rat neural cell cultures. In Vitro Toxicology, 9, 361-371.

105. Monticello, T.M., Swenberg, J.A., Gross, E.A., Leininger, J.R., Kimbell, J.S., Seilkop, S.K., Starr, T.B., Gibson, J.E. and Morgan, K.T. (1996). Correlation of regional and nonlinear formaldehyde-induced cancer with proliferating populations of cells. Cancer Res, 56, 1012-1022.

106. Templin, M.V., Jamison, K.C., Wolf, D.C., Morgan, K.T., and Butterworth, B.E. (1996). Comparison of chloroform-induced toxicity in the kidneys, liver, and nasal passages of male Osborne-Mendel and F-344 rats. Cancer Lett., 104, 71-78.

107. Owens, J.G., James, R.A., Moss, O.R., Morgan, K.T., Bowman, J.R., Struve, M.F. and Dorman, D.C. (1996). Design and evaluation of an olfactometer for the assessment of 3-methylindole-induced Hyposmia. Fundam. Appl. Toxicol., 32, In press.

108. Kimbell, J.S., Gross, E.A., Richardson, R.B., Conolly, R.B., and Morgan, K.T. (1996). Correlation of regional formaldehyde flux predictions with the distribution of formaldehyde-induced squamous metaplasia in F344 rat nasal passages. Mutation Res. In Press.

109. Ibanes, J.D., Leininger, J.R., Jarabek, A.M., Harkema, J.R., Hotchkiss, J.A. and Morgan, K.T. (1996). Re-examination of respiratory tract responses in rats, mice and rhesus monkeys chronically exposed to inhaled chlorine. Inhal. Toxicol., 8:859-876.

110. Dorman, D.C., Struve, M.F., Wong, B.A., Morgan, K.T., Janszen, D., Gross, E.A., and Bond, J.A. (1997). Neurotoxicologic

evaluation of ethyl tertiary butyl ether inhalation in the Fischer 344 rat. J. Appl. Toxicol., 17, 235-242.

111. Dorman, D.C., Miller, K.L., D'Antonio, A., James, R.A., and Morgan, K.T. (1997). Chloroform-induced olfactory mucosal degeneration and osseous ethmoid hyperplasia are not associated with olfactory deficits in Fischer 344 rats. Toxicology, 122, 39-50.

112. Morgan, K.T. (1997). A brief review of formaldehyde carcinogenesis in relation to rat nasal pathology and human health risk assessment. Toxicol. Pathol., Invited Review, 25:291-307.

113. Kepler, G.M., Richardson, R.B., Morgan, K.T. and Kimbell, J.S. (1998). Computer simulation of inspiratory nasal airflow and inhaled gas uptake in a rhesus monkey. Toxicol. Appl. Pharmacol., 150:1-11.

114. Frederick, C.B., Lomax, L.G., Black, K.A., Finch, L., Bush, M.L., Ultman, J.S., Kimbell, J.S., Morgan, K.T., Subramaniam, R.P., Morris, J.B., Stott, W.T., Young, J.T. and Scherer, P.W. Application of computational fluid dynamics and a physiologically-based inhalation model for interspecies extrapolation of the dosimetry of acidic vapors in the upper respiratory tract. Toxicol. Appl. Pharmacol., .

115. Crosby, L.M., Morgan, K.T., Gaskill, B., Wolf, D.C. and DeAngelo, (2000). D.B. Origin and distribution of potassium bromate-induced testicular and peritoneal mesotheliomas. Toxicol. Pathol., 28:253-266.

116. Kimbell, J.S., Subramaniam, R.P., Gross, E.A., Schlosser, P.M., Georgieva, A., Gilstrap, C.L. and Morgan, K.T. (1999). Predictions of inhaled formaldehyde gas uptake in the rat, monkey, and human nasal passages. Published.

117. Kepler, T.B., Crosby, L.M. and Morgan, K.T. (2000). Normalization and analysis of DNA microarray data by self-consistency and local regression, Genome Biology, 3, 1-12.

118. Crosby, L.M., Hyder, K.S., DeAngelo, A.B., Kepler, T.B., Gaskill, B., Benavides, G.R., Yoon, L. and Morgan, K.T. (2000). Morphologic analysis correlates with gene expression changes in

cultured F344 rat mesothelial cells. Toxicol. Appl. Pharmacol., 169, 205-221.

119. Kimbell, J.S., Subramaniam, R.P., Gross, E.A., Schlosser, P.M. and Morgan, K.T. (2001). Comparison of inhaled formaldehyde gas uptake predictions in the rat, monkey, and human nasal passages. Toxicol. Sci., .

120. Kimbell, J.S., Overton, J.H., Subramaniam, R.P., Schlosser, P.M., Morgan, K.T., Conolly, R.B. and Miller, F.J. (2001). Simulation of local formaldehyde flux in rat, monkey, and human nasal passages for health risk assessment. Toxicol. Sci, .

121. Morgan, K.T., Brown, H.R., Benavides, G., Crosby, L., Sprenger, D., Yoon, L., Ni, H., Easton, M., Morgan, D., Morgan, D., Laskowitz, D. and Tyler, R. (2002). Toxicogenomics and human disease risk assessment. HERA, 8, 1339-1353.

122. Anderson, M., Bogdanffy, M., David Dankovic, Elaine Faustman, Paul Foster, Clay Frederick, Barry Johnson, Carole Kimmel, Steven Lewis, Roger McClellan, Ron Melnick, Frank Mirer, Kevin Morgan, Val Schaeffer, Ellen Silbergeld, William Slikker, James Swenberg, Mark Toraason, Harri Vainio, Elizabeth Ward (2000). Improving Experimental Toxicological Research for Risk Assessment. HERA, In Press.

123. Morgan, K.T., Hong Ni, H. Roger Brown, Lawrence Yoon, Charles W. Qualls, Jr., Lynn M. Crosby, Randall Reynolds, Betty Gaskill, Steven P. Anderson, Tom B. Kepler, Tracy Brainard, Nik Liv, Marilyn Easton, Christine Merrill, Don Creech, Dirk Sprenger, Gary Conner, Paul R. Johnson, Tony Fox, Ron Tyler, Maureen Sartor, Erika Richard, Sabu Kuruvilla, Warren Casey, Gina Benavides.(2002). Application of cDNA Microarray Technology to In Vitro Toxicology and the Selection of Genes for a Real Time RT-PCR-Based Screen for Oxidative Stress in Hep-G2 Cells. Toxicol. Pathol., 40: 435.

124. Morgan, K.T., Casey, W., Easton, M., Creech, D., Ni, H., Yoon, L., Anderson, S., Qualls, C.W.Jr., Crosby, L.M., MacPherson, A., Bloomfield, P., and Elston, T.C. (2003). Frequent sampling reveals dynamic responses by the transcriptome to

routine media replacement in HepG2 cells. Toxicol. Pathol., 31, 448-461.

125. Merrill, C.L., Ni, H., Yoon, L.W., Tirmenstein, M.A., Narayanan, P., Benavides, G.R., Easton, M.J., Creech, D.R., Hu, C.X., McFarland, D.C., Hahn, L.M., Thomas, H.C. and Morgan, K.T. (2002). Etomoxir-induced oxidative stress in HepG2 cells detected by differential gene expression is confirmed biochemically. Toxicol. Sci., 69, 93-101.

126. Boorman, G.A., Anderson, S.P., Casey, W.M., Brown, H.R., Crosby, L.M., Gottschalk, K., Easton, M., Ni, H., and Morgan, K.T. (2002). Toxicol. Pathol., 30, 15-27.

127. Casey, W., Anderson, S., Fox, T., Dold, K., Colton, H. and Morgan, K.T. (2002). Transcriptional and physiological responses of HepG2 cells exposed to diethyl maleate: time course analysis. Physiol. Genomics, 8, 115-122.

128. Brown, H.R., Ni, H., Benavides, G., Yoon, L., Hyder, K., Giridhar, J., Gardiner, G., Tyler, R.D. and Morgan, K.T. (2002). Correlation of simultaneous differential gen expression in the blood and heart with known mechanisms of adriamycin-induced cardiomyopathy in the rat. Toxicol. Pathol., 30, 452-469.

129. Kuruvilla, S., Qualls, C.W.Jr., Tyler, R.D.,Witherspoon, S.M., Benavides, G.R., Yoon, L.W., Dold, K., Brown, H.R., Sangiah, S., and Morgan, K.T. (2003). Effects of minimually toxic levels of carbonyl cyanide P-(trifluoromethoxy) phenylhydrazone (FCCP), elucidated through differential gene expression with biochemical and morphological correlations. Toxicol. Sci., 73, 348-361.

130. Morgan, K.T., Pino, M., Crosby, L.M., Wang, M., Elston, T.C., Jayyosi, Z., Bonnefoi, M. and Boorman, G. (2004). Complementary roles for toxicologic pathology and mathematics in toxicogenomics, with special reference to data interpretation and oscillatory dynamics. Toxicol. Pathol., 32 (suppl.), 13-25.

131. Morgan, K.T., Jayosi, Z., Hower, M.A., Pino, M.V., Connolly, T.M., Kotlenga, K., Lin, J., Wang, M.,Schmidts, H-L, Bonnefoi, M.S., Elston, T.C. and Boorman, G.A. (2005). The

hepatic transcriptome as a window on whole-body physiology and pathophysiology. Toxicol. Pathol., 33, 136-145.

132. Kim, Y., Ton, T.V., DeAngelo, A.B., Morgan, K.T., Devereaux, T.R., Anna, C., Collins, J.B., Paules, R.S., Crosby, L. M., and Sills, R. (2006). Major carcinogenic pathways identified by gene expression analysis of peritoneal mesotheliomas following chemical treatment in F344 rats.

133. Ke Xu, Kevin T. Morgan, Abby *Todd* Gehris, Timothy C. Elston , Shawn M. Gomez (2011). A Whole-Body Model for Glycogen Regulation Reveals a Critical Role for Substrate Cycling in Maintaining Blood Glucose Homeostasis. PLOS Computational Biology, December 1, 2011 DOI: 10.1371/journal.pcbi.1002272

ABOUT THE AUTHOR

Kevin Thomas Morgan is a retired veterinary pathologist and research scientist. He now works on ways to help older people keep going to enjoy every day they are lucky enough to have. He does this by writing books, creating instructional videos, and giving inspiring talks to groups of seniors. His current interests include reading and learning to write. He enjoys solving problems to help people in pain. Kevin is an avid Ironman-distance triathlete, vegetable gardener and vegan. Some of his work is designed to help people who like himself have aortic and other vascular diseases. He enjoys friends, family and not being dead for as long as possible.

NEWSLETTER SIGN UP LINK

Old Dogs in Training LLC newsletters are published every now and then. They relate to preparing for aging, living with vascular disease, including aortic aneurysms, safe exercise for older people, nociceptive foot pain aka plantar fasciitis, and other subjects that interest the author with respect to enjoying our later years.

You can sign up for the newsletter on the authors blog, *AthleteWithStent.Com*.

Copyright © 2018

Kevin Thomas Morgan, Old Dogs in Training, LLC.

All rights reserved. This book may not be reproduced, in whole or in part, in any form or by any means electronic or mechanical, including photocopying, recording or by any information storage and retrieval system now known or hereafter invented, without written permission from the author, Kevin Thomas Morgan, aka FitOldDog.

Copyrights for images were purchased from, and are filed with, ShutterStock, Inc. Mockups were created with tools purchased from MediaModifier, Inc.

LIMIT OF LIABILITY AND DISCLAIMER

This document is based on the author's experience and has been created to provide information and guidance about the subject matter covered. Every effort has been made to make it as accurate as possible. Website links and content can change at any time. If a link is nonfunctional, please contact the author. The author shall have neither liability nor responsibility to any person or entity with respect to any loss, damage or injury caused or alleged to be caused directly or indirectly by the information covered in this book, or by decisions made based upon this book.

TRADEMARKS

Any trademarks, service marks, product names or named features are assumed to be the property of their respective owners, and are used for reference only.

SHARING THIS DOCUMENT

I politely ask that you please respect my work by not donating or reselling this book. This would be very much appreciated.

MEDICAL DISCLAIMER

As a veterinarian, I do not provide medical advice to human animals. If you undertake or modify an exercise program, consult your medical advisers before doing so. Undertaking activities pursued by the author does not mean that he endorses your undertaking such activities, which is clearly your decision and responsibility. Be careful and sensible, please. Old Dogs in Training, LLC.

❦ Created with Vellum

www.ingramcontent.com/pod-product-compliance
Lightning Source LLC
Chambersburg PA
CBHW072139170526
45158CB00004BA/1434